MONOGRAPHS ON
APPLIED PROBABILITY AND STATISTICS

DISTRIBUTION-FREE STATISTICAL METHODS

MONOGRAPHS ON
APPLIED PROBABILITY AND STATISTICS

General Editor
D.R. COX, F.R.S.

Also available in the series

Probability, Statistics and Time
M.S. Bartlett

The Statistical Analysis of Spatial Pattern
M.S. Bartlett

Stochastic Population Models in Ecology and Epidemiology
M.S. Bartlett

Risk Theory
R.E. Beard, T. Pentikäinen and E. Pesonen

Point Processes
D.R. Cox and V. Isham

Analysis of Binary Data
D.R. Cox

The Statistical Analysis of Series of Events
D.R. Cox and P.A.W. Lewis

Queues
D.R. Cox and W.L. Smith

Stochastic Abundance Models
E. Engen

The Analysis of Contingency Tables
B.S. Everitt

Distribution-Free Statistical Methods

J.S. MARITZ

Department of Mathematical Statistics
La Trobe University
Australia

LONDON   NEW YORK

CHAPMAN AND HALL

First published 1981 by
Chapman and Hall Ltd
11 New Fetter Lane, London EC4P 4EE

Published in the USA by
Chapman and Hall
in association with Methuen, Inc.
733 Third Avenue, New York, NY 10017

Printed in Great Britain at the
University Press, Cambridge

British Library Cataloguing in Publication Data

Maritz, J. S.
Distribution-free statistical methods. – (Monographs on
applied probability and statistics)
1. Probabilities   2. Mathematical statistics
I. Title   II. Series
519.2   QA278   80–42136
·M36 1981
ISBN 0–412–15940–6

Contents

Preface

The preparation of several short courses on distribution-free statistical methods for students at third and fourth year level in Australian universities led to the writing of this book. My criteria for the courses were, firstly, that the subject should have a clearly recognizable underlying common thread rather than appear to be a collection of isolated techniques. Secondly, some discussion of efficiency seemed essential, at a level where the students could appreciate the reasons for the types of calculations that are performed, and be able actually to do some of them. Thirdly, it seemed desirable to emphasize point and interval estimation rather more strongly than is the case in many of the fairly elementary books in this field.

Randomization, or permutation, is the fundamental idea that connects almost all of the methods discussed in this book. Application of randomization techniques to original observations, or simple transformations of the observations, leads generally to conditionally distribution-free inference. Certain transformations, notably 'sign' and 'rank' transformations may lead to unconditionally distribution-free inference. An attendant advantage is that useful tabulations of null distributions of test statistics can be produced.

In my experience students find the notion of asymptotic relative efficiency of testing difficult. Therefore it seemed worthwhile to give a rather informal introduction to the relevant ideas and to concentrate on the Pitman 'efficacy' as a measure of efficiency.

Most of the impetus to use distribution-free methods was originally in hypothesis testing. It is now well recognized that adaptation of some of the ideas to point estimation can be advantageous from the points of view of efficiency and robustness. Pedagogically there are also advantages in emphasizing estimation. One of them is that one can adopt the straightforward approach of defining relative efficiency in terms of variances of estimates. Another is that using the notion of an estimating equation makes it easy to relate the distribution-free techniques to methods which will have been encountered in the standard statistics courses. Examples include the method of moments, and large sample approximations to standard errors of estimates.

The aim of this book is to give an introduction to the distribution-free way of thinking and sufficient detail of some standard techniques to be useful as a practical guide. It is not intended as a compendium of distribution-free techniques and some readers may find that a technique which they regard as important is not mentioned. For the most part the book deals with problems of location and location shift. They include one- and two-sample location problems, and some aspects of regression and of the 'analysis of variance'.

Although some of the presentation is somewhat different from what appears to have become the standard in this field, very little, if any, of the material is original. Much has been gleaned from various texts. Direct acknowledgement of my indebtedness to the authors of these works is made by the listing of general references. Through these, and other references, I also acknowledge indirectly the work of other authors whose names may not appear in the bibliography. No serious attempt has been made to attribute ideas to their originators. Specific references are given only where it is felt that readers may be particularly interested in more detail.

While the origins of this book are in undergraduate teaching I do hope that some experienced statisticians will find parts of it interesting. In particular, developments in point and interval estimation, and noting of their connections with 'robust' methods have taken place fairly recently. Some interesting problems of estimating standard errors, as yet not fully resolved, are touched upon in several places.

Many of my colleagues have helped me, by discussion and by reading sections of manuscript. Dr D.G. Kildea read the first draft of Chapter 2 and his detailed comments led to many improvements. Dr B.M. Brown was not only a patient listener on many occasions but also generously provided Appendix A.

Melbourne, November 1980   J.S. Maritz

CHAPTER 1

Basic concepts in distribution-free methods

1.1 Introduction

In the broadest sense a distribution-free statistical method is one that does not rely for its validity or its utility on any assumptions about the form of distribution that is taken to have generated the sample values on the basis of which inferences about the population distribution are to be made. Obviously a method cannot be useful unless it is valid, but the converse is not true. The terms validity and utility are used in a semi-technical sense and relate to the usual statistical notions of consistency and efficiency respectively. The great attractions of distribution-free methods are:

(i) that they are, by definition, valid under minimal assumptions about underlying distributional forms;

(ii) the aesthetic appeal of their being based for the most part on very simple permutation or randomization ideas;

(iii) the fact that they have very satisfactory efficiency and robustness properties.

Distribution-free methods, especially the simpler ones, have gained widespread acceptance, but they are by no means the first weaponry of most practising statisticians. Perhaps the main impediments to their even greater popularity are:

(a) the results of distribution-free tests are often not as readily interpretable in terms of physical quantities as are the results of parametric analysis;

(b) in some of the more complex situations severe computational difficulties can arise; although many distribution-free methods are 'quick' and 'easy' they do not all share these properties.

It should also be noted, of course, that many distribution-free methods are relatively new; this applies particularly to the estimation methods. Therefore they are not yet well known in the popular sense.

This book is written for undergraduate students, as an instruction manual in the use of some of the standard distribution-free methods, but its main objective is to serve as an introduction to the underlying ideas, and perhaps to stimulate further reading in the subject. Consequently the emphasis is on randomization (or permutation) as the underlying unifying notion in the development of testing methods, and associated methods of estimation. These ideas are certainly not new, and have been developed in great detail in various special contexts. Nevertheless, the use of 'signs' and 'ranks' is still commonly thought to characterize distribution-free methods, if not by statisticians, then by very many non-professional users of statistical methods.

The selection of topics that are treated in ensuing chapters is influenced strongly by consideration (a) above. In fact the emphasis is heavily on questions of location and location shift. They are not only among the most important from the practical viewpoint, but also represent a class of problems where it is clearly easy, and sensible, to visualize the quantities that are subject to inference, without the need to specify underlying distributions in close detail. This is the only excuse offered for not including many 'standard' procedures, such as runs tests, some of the tests of dispersion, general tests of distribution functions, such as the Kolmogorov–Smirnov test.

Very few of the so-called distribution-free methods are truly distribution-free. Many of the arguments are simplified if the underlying distribution can be taken as continuous, and this is commonly done. This assumption will be made throughout this book. Other assumptions are necessary, depending on the problem. For example, in one-sample location problems the assumption of symmetry plays a major role. Thus the term distribution-free must be interpreted with some qualification. The methods are developed without detailed parametric specification of distributions; we may assume that a density $f(x)$ is symmetric about θ, but need not say that it is, for example, $1/[\pi\{1 + (x - \theta)^2\}]$. The term 'nonparametric' is preferred to 'distribution-free' by some, but since we are actually trying to make inferences about parameters the latter term seems more appropriate here.

1.2 Randomization and exact tests

Although the randomization basis of test and other methods will be restated for specific procedures in later chapters, we shall illustrate it here in a simple example. This will enable us to define certain terms rather conveniently.

Consider the well-known simple 'paired comparison' experiment in which two treatments are allocated at random, one each to a pair of subjects. The two subjects of the same pairs are chosen to be as alike as possible. For example, in an experiment on sheep a natural pairing would be to use twins for each pair. Suppose that the results obtained for the ith pair are measurements y_{iA} and y_{iB} for the members receiving treatments A and B respectively. The differential effect of the treatments for the ith pair could now be measured by $d_i = y_{iA} - y_{iB}$, with $i = 1, 2, \ldots, n$.

Now suppose that we are to test the null hypothesis H_0, according to which the effects of treatments A and B are identical. Since our allocation of treatments within pairs is random, correctness of H_0 would mean that d_i could have had the value $-d_i = y_{iB} - y_{iA}$. Further, if we denote by D_i the random variable being the ith difference obtained in the experiment, then

$$\Pr(D_i = +|d_i|) = \Pr(D_i = -|d_i|) = \tfrac{1}{2} \tag{1.1}$$

The probabilities in (1.1) are conditional probabilities, the conditioning being on the ith pair whose difference has magnitude $|d_i|$.

Since the randomization is performed independently for each pair, it is now a simple matter to conceive the joint distribution of $D_1, D_2, \ldots, D_n$, for the random variables $D_1, D_2, \ldots, D_n$ are independent with individual distributions given by (1.1). Again we note that it is a conditional distribution. A natural test statistic for H_0 is $T = D_1 + D_2 + \ldots + D_n$, and from the preceding discussion it is clear that tabulation of the exact conditional distribution of T is a straightforward matter; the 2^n possible sign combinations to be attached to the magnitudes $|d_i|$, $i = 1, 2, \ldots, n$, have to be listed, and for each of these the value of T computed. This generates 2^n possible values of T occurring with equal probabilities 2^{-n}, and thus establishes the exact conditional distribution of T. Let τ be such that $\Pr(T \geqslant \tau) = r/2^n$. If we test H against a one-sided alternative and take as critical region all $T_i \geqslant \tau$, then the size of this critical region (the level of significance) is exactly $r/2^n$.

A test is said to be *exact* if the actual significance level is exactly that which is nominated. In our example, the test is an exact level $r/2^n$ test. Moreover, it is important to note that although the exact significance level derives from the exact conditional distribution, the unconditional significance level is also exactly $r/2^n$. This is true simply because, for every possible set $|d_i|$, $i = 1, 2, \ldots n$, the probability of rejecting H_0 is $r/2^n$.

The distribution of T under H_0 will often be referred to as the null distribution of T. In our example it is a conditional null distribution. The test of H_0 is carried out by referring the observed value of T to its conditional null distribution. The derivation of the distribution in our example was achieved by using the randomization argument, and did not depend on an assumption of the form of distribution that individual y_i values might follow. So, the conditional null distribution of T does not depend on an underlying distributional assumption, and consequently the significance level is exact, free from such an assumption.

Although the exactness of the test is not affected by its being based on a conditional null distribution, it should be kept in mind that the conditional distribution of T will change from sample to sample. Therefore the unconditional distribution of T will be a mixture of the conditional distributions and its form will depend on the underlying distribution of $|d|$ values. When the exact conditional distribution of a statistic, obtained by a randomization procedure, is not invariant with respect to the realized sample values, the associated distribution-free methods are said to be *conditionally distribution-free*.

By transformations such as rank and sign transformations it is often possible to derive methods that are unconditionally distribution-free from those that are conditionally distribution-free. If every $|d_i|$ in the example that we have been discussing is replaced by 1 we obtain the well known 'sign test' and it is clear that the conditional distribution of $S = \sum_{i=1}^{n} \text{sgn}(D_i)$, remains exactly the same for every possible set of realized results.

From the point of view of exactness of significance levels, there is no obvious advantage in a test being unconditionally distribution-free. Since the distributions of test statistics can be tabulated once and for all if they are unconditionally distribution-free, there can be worthwhile computational advantages in such tests. We shall see, also, that there can be gains in efficiency by astute choice of transformation. However, our starting point is randomization and its natural consequence is to produce, in the first instance, conditionally distribution-free methods.

Enumeration of exact null distributions can be a totally impractical task for large sample sizes, hence it is quite common to approximate null distributions by some standard distribution, usually a normal distribution, and so to obtain approximate values of significance levels. Here the approximation is a mathematical convenience and does not affect the exactness of the method in principle. Hence such procedures will be called 'exact' whether or not some mathematical approximation is used for convenience. However, there are circumstances where approximations, of essentially unknown precision, have to be made. They are usually occasioned by nuisance parameters, whose values, while not of direct concern do affect the null distributions of interest.

1.3 Consistency of tests

If the null hypothesis in the example of Section 1.2 had specified a difference θ between the two treatments, the d_i values would have been replaced by $d_i - \theta$, $i = 1, 2, \ldots, n$, the argument otherwise remaining the same. In order to show the dependence of T on $D_1, D_2, \ldots, D_n$ and θ we may write it $T(\mathbf{D}, \theta)$. Taking differences of the paired values can be regarded as reducing the problem to a one-sample problem, and in the remainder of this section we shall be discussing one-sample, one-parameter problems.

Suppose that random sampling from a population with parameter of interest θ produces the results $X_1, X_2, \ldots, X_n$, where $X_1, X_2, \ldots, X_n$ can be taken as independent and identically distributed. Let the statistic to be used in testing a hypothesis about θ be $S(\mathbf{X}, t)$, defined such that its conditional and its unconditional null distributions have mean 0 if t is replaced by θ.

Suppose that we propose to test $H_0 : \theta = \theta_0$ against $H_1 : \theta = \theta_1 > 0$ at level α and that the test procedure is to reject H_0 if observed $S(\mathbf{X}, \theta_0) > C_\alpha(\mathbf{X}, \theta_0)$. The value of $C_\alpha(\mathbf{X}, \theta_0)$ is determined from the randomization distribution of $S(\mathbf{X}, \theta_0)$, therefore it generally depends on $\mathbf{X}$ and on θ_0. We shall assume that $S(\mathbf{X}, \theta_0)$ is so scaled with respect to n that $C_\alpha(\mathbf{X}, \theta_0) \xrightarrow{P} 0$, i.e., in probability, as $n \to \infty$.

Questions of consistency have to be answered in terms of the behaviour of the unconditional distributions of the relevant statistics, and we shall assume that in the unconditional distribution of $S(\mathbf{X}, \theta_0)$, when $\theta = \theta_1$,

$$E\{S(\mathbf{X}, \theta_0)|\theta_1)\} = \Delta(\theta_1, \theta_0) > 0$$
$$\operatorname{var}\{S(\mathbf{X}, \theta_0)|\theta_1)\} = \sigma^2(\theta_1, \theta_0)/n, \text{ with } \sigma^2(\theta_1, \theta_0) \text{ bounded.}$$

We shall say that the test of H_0 against H_1 is *consistent* if its power can be made arbitrarily close to 1 by increasing n.

Lemma 1.1 Under the assumptions given above about the distribution of $S(\mathbf{X}, t)$, the test of H_0 against H_1 is consistent.

The proof of this lemma is obtained simply from the assumptions by noting that $S(\mathbf{X}, \theta_0) \xrightarrow{P} \Delta(\theta_1, \theta_0) > 0$, while $C_\alpha(\mathbf{X}, \theta_0) \xrightarrow{P} 0$.

We may be concerned with values of θ_1 close to θ_0 in which case it is useful if we assume

$$E\{S(\mathbf{X}, t)|\theta\} = \mu(t, \theta)$$

with $\mu(t, \theta)$ continuous in t and differentiable near θ.
Then we can put

$$\Delta(\theta_1, \theta_0) \simeq (\theta_0 - \theta_1)\mu'(\theta_1, \theta_1),$$

noting that $\mu(\theta_0, \theta_0) = 0$.

For Lemma 1.1 to hold we now have to replace the earlier assumption about $E\{S(\mathbf{X}, \theta_0)|\theta_1\}$ by the assumption $\mu'(\theta, \theta) < 0$.

Let us reconsider the example of Section 1.2 in two versions.

(i) $S(\mathbf{X}, t) = T^*(\mathbf{D}, t) = T(\mathbf{D}, t)/n = (D_1 + \ldots + D_n)/n - t$; suppose, for our illustration, that the distribution of every D_i is normal with mean θ and variance σ^2, and that $\theta_0 = 0$. In this case, as we shall see, the conditional randomization distribution of S can be taken as approximately normal for large n, with variance $\sum d_i^2/n^2$, if $d_1, d_2, \ldots, d_n$ are the observed differences. So, with u_α an appropriate normal quantile,

$$C_\alpha(\mathbf{X}, \theta_0) = u_\alpha\{\sum D_i^2/n\}^{1/2}/\sqrt{n},$$

and it is easy to see that $\sum D_i^2/n \xrightarrow{P} \sigma^2 + \theta_1^2$ under H_1 so that $C_\alpha(\mathbf{X}, \theta_0) \xrightarrow{P} 0$.

Further,

$$E\{S(\mathbf{X}, \theta_0)|\theta_1\} = \theta_1$$
$$\mathrm{var}\,\{S(\mathbf{X}_1, \theta_0)|\theta_1\} = \sigma^2/n$$

note also that $E\{S(\mathbf{X}, t)|\theta\} = \theta - t$, $\mu'(\theta, \theta) = -1$. The test is consistent; in fact, as we shall show in Chapter 2 the test statistic can be written as a function of the usual t-statistic whose consistency and other properties are well known.

It should be noted, however, that with certain distributions for the D_i the test may not be consistent.

(ii) Let $S(\mathbf{X}, t) = (1/n)\sum_{i=1}^{n} \mathrm{sgn}(D_i - t)$ and suppose that the distribution function of every D_i is $F(d, \theta)$, with density $f(d, \theta)$, symmetric about θ. The null distribution of S has variance $1/n$, n even, and S, being a linear function of a binomial random variable, has an approximately normal distribution. So,

$$C_\alpha(\mathbf{X}, \theta_0) = u_\alpha/\sqrt{n}$$

Also

$$\begin{aligned}
E\{S(\mathbf{X}, t)|\theta\} &= (1/n) \sum E\{\text{sgn}(D_i - t)|\theta\} \\
&= (1/n) \sum \{(-1) P(D_i < t|\theta) + (1) P(D_i > t|\theta)\} \\
&= 1 - 2F(t, \theta)
\end{aligned}$$

Therefore $\mu'(\theta, \theta) = -2f(\theta, \theta)$. Under H_1 the statistic S can again be expressed as a binomial random variable, hence $\sigma^2(\theta_1, \theta_0) \leqslant \frac{1}{4}$, so that the conditions of Lemma 1.1 are satisfied.

Two points about the examples above are worth remarking upon. First we see that the quantity $C_\alpha(\mathbf{X}, \theta_0)$ is simply a constant, that is non-random, in case (ii). Second, the only restriction on F is that $f(\theta, \theta) \neq 0$, so that the 'sign' test will be consistent in many cases where the usual t-test is not.

1.4 Point estimation of a single parameter

Since the statistic $S(\mathbf{X}, t)$ is defined so that $E\{S(\mathbf{X}, \theta)\} = 0$, a natural procedure for finding a point estimate of θ is suggested by the method of moments, namely, to take as point estimate $\hat{\theta}$ of θ the solution $t = \hat{\theta}$ of the estimating equation

$$S(\mathbf{X}, t) = 0. \tag{1.2}$$

As we shall see in later chapters, some of the statistics $S(\mathbf{X}, t)$, regarded as functions of t for fixed $\mathbf{X}$ are not continuous in t so that a unique solution of (1.2) has to be decided upon by a suitable convention. An appropriate one is usually obvious in context. Weak consistency of $\hat{\theta}$ is easily checked; in fact we have:

Lemma 1.2 If $E\{S(\mathbf{X}, t)|\theta\} = \mu(t, \theta)$ is continuous in t and differentiable near $t > \theta$, and if the other conditions for the applicability of Lemma 1.1 hold, then $\hat{\theta} \xrightarrow{P} \theta$ as $n \to \infty$.

1.5 Confidence limits

Provision of a measure of precision of an estimate is an essential part of statistical inference, and one way of doing this is to give confidence limits. Confidence limits, or confidence sets, can be determined by the well-known procedure of taking a point θ' in the parameter space to belong to the confidence set if the null hypothesis that $\theta = \theta'$ is accepted. Briefly, the argument is as follows:

To test $H_0 : \theta = \theta_0$ against some alternative H_1, a critical region of size α, $R(\theta_0)$, is determined such that

$$\Pr\{S(\mathbf{X}, \theta_0) \in R(\theta_0) | H_0\} = 1 - \alpha \tag{1.3}$$

Now, for a given $\mathbf{X} = \mathbf{x}$ find the set $C_\theta(\mathbf{x})$ of all θ such that $S(\mathbf{x}, \theta) \in R(\theta)$. The true value θ_0 will belong to $C_\theta(\mathbf{x})$ if $S(\mathbf{x}, \theta_0) \in R(\theta_0)$. But the probability of this event is $1 - \alpha$, whatever the value of θ_0, according to our definition (1.3). The set $C_\theta(\mathbf{x})$ is a $100(1 - \alpha)\%$ confidence set for θ. The shape of $C_\theta(\mathbf{x})$ is determined by the shape of $R(\theta)$. In the one-parameter case that we are considering, $R(\theta)$ is typically an interval, and so is $C_\theta(\mathbf{x})$; one- or two-sided confidence limits are obtained according to whether the test is one- or two-sided. The probability $1 - \alpha$ is sometimes called the *confidence coefficient*.

A notable feature of the procedure outlined above is that only the null distribution of S is needed. In distribution-free methods this is particularly useful because the null distributions are usually exact, often very easy to obtain, and of course in many instances already tabulated. Moreover, whether a conditional or unconditional null distribution is used the confidence coefficient is the value $1 - \alpha$ associated with the hypothesis test, and if the probability $1 - \alpha$ is exact, then so is the confidence coefficient. We shall say that a confidence region is exact if the confidence coefficient $1 - \alpha$ is exact. One of the great attractions of distribution-free methods is that they enable one to determine, often fairly easily, exact confidence limits for certain parameters with minimal assumptions about distributional forms.

1.6 Efficiency considerations in the one-parameter case

1.6.1 *Estimation*

Efficiency of estimation will be measured in terms of $\text{var}(\hat{\theta})$. The relative efficiency of two estimators will be measured by the ratio of their variances. In some cases it will be possible to express $\hat{\theta}$ fairly simply in terms of $X_1, X_2, \ldots, X_n$ so that an exact expression for $\text{var}(\hat{\theta})$ may be given. However, for the most part we shall have to deal with cases where such a simple expression cannot be obtained; in fact we may not even be able to express $\hat{\theta}$ explicitly in terms of $X_1, X_2, \ldots, X_n$. Then the best we can do is obtain a large-sample approximation formula for $\text{var}(\hat{\theta})$.

To simplify notation we shall assume in what follows that expectation and variance are derived at the true value of the parameter θ. Thus we put $E\{S(\mathbf{X}, t)|\theta\} = E\{S(\mathbf{X}, t)\}$, etc. The main assumptions that we shall make are:

(i) $E\{S(\mathbf{X}, t)\} = \mu(t, \theta)$, is continuous in t and differentiable near $t = \theta$;

(ii) the statistic $S(\mathbf{X}, t)$, treated as a function of t for fixed $\mathbf{X}$, either is continuous and differentiable for t near θ, or it can be replaced by an approximating function which has these properties;

(iii) $\text{var}\{S(\mathbf{X}, t)\} = \sigma^2(t, \theta)/n$, $\sigma^2(t, \theta)$ continuous in t and bounded.

Assumption (ii) is needed because statistics $S(\mathbf{X}, t)$ obtained after rank or sign transformations are typically discontinuous step functions of t. However, it is also typically true of them that if they are scaled such that $E\{S(\mathbf{X}, t)\} = \mu(t, \theta)$, that is, not dependent on n, then the number of steps increase with n, and their heights decrease. The sign statistic of example (ii) in Section 1.3 is a case in point; a simple transformation of $S(\mathbf{X}, t)$ is the sample distribution function which is known to have the desired property.

Now consider a small but finite neighbourhood of θ, the interval $(\theta - h/2, \theta + h/2)$ with h held constant, so that we can put

$$\left\{ \frac{\partial S(\mathbf{X}, t)}{\partial t} \right\}_{t = \theta} \simeq \frac{S(\mathbf{X}, \theta + h/2) - S(\mathbf{X}, \theta - h/2)}{h}$$

$$= \frac{S(\theta + h/2, \theta) - S(\theta - h/2, \theta) + O(1/\sqrt{n})}{h}$$

$$\simeq \mu'(\theta, \theta) + O(1/\sqrt{n})$$

in view of assumption (iii).

Write

$$S(\mathbf{X}, t) = S(\mathbf{X}, \theta) + (t - \theta)\{\mu'(\theta, \theta) + O(1/\sqrt{n})\} \qquad (1.4)$$

and note that $S(\mathbf{X}, \hat{\theta}) = 0$. Then we have approximately

$$\text{var}(\hat{\theta}) = \text{var}\{S(\mathbf{X}, \theta)\}/[\partial E\{S\mathbf{X}, t)\}/\partial t]_{t = \theta}^2 \qquad (1.5)$$

Formula (1.5), or approximate methods similar to those used in its derivation, occur in sundry standard situations. For example, if it is applied to the estimating equation in the case of 'regular' maximum-likelihood estimation, the usual large-sample variance formula for maximum-likelihood estimators is obtained.

A simple example is the following: suppose that $F_n(x)$ is the usual sample distribution function based on n independent observations from a population with continuous distribution function $F(x)$, density $f(x)$ and median θ. The estimating equation for θ is

$$S(\mathbf{X}, t) = F_n(t) - 1/2 = 0$$
$$E\{S(X, t)\} = F(t) - 1/2$$
$$[\partial E\{S(\mathbf{X}, t)\}/\partial t]_{t=\theta} = r(\theta)$$
$$\text{var}\{S(\mathbf{X}, \theta)\} = 1/4n$$

giving the approximate formula for the variance of the sample median, $\hat{\theta}$,

$$\text{var}(\hat{\theta}) \simeq 1/(4nf^2(\theta)).$$

Another consequence of the 'linearization' represented by (1.3) is that the distribution of $\hat{\theta}$ will be approximately normal if the distribution of $S(\mathbf{X}, \theta)$ is approximately normal. Linearization ideas have become important in distribution-free methods, and a more rigorous approach is outlined in Appendix A.

1.6.2 Hypothesis testing

Consider two statistics S_1 and S_2 satisfying the conditions (i), (ii), (iii) given in Section 1.6.1. Suppose that we scale these two statistics so that their null distributions have the same dispersion at a certain value θ_0 of θ; that is, we replace S_1 by $S_1(\mathbf{X}, t)/\sigma_1(\theta_0, \theta_0) = S_1^*(X, t)$ and $S_2(X, t)$ by $S_2(X, t)/\sigma_2(\theta_0, \theta_0) = S_2^*(\mathbf{X}, t)$. Inspection of formula (1.4) shows that the ratio of variances of estimates of θ_0 based on S_1^* and S_2^* is determined by the slopes $[\partial E\{S_r^*(\mathbf{X}, t)\}/\partial t]_{t=\theta_0}$, $r = 1, 2$.

Without doing formal power calculations, it is clear that the power of a test of $H_0 : \theta = \theta_0$ against $H_1 : \theta = \theta_0 + \Delta$, Δ small, will be determined largely by the slope given above associated with the statistic S_r^* used for the test. For any statistic S, the slope at θ_0,

$$e_S(\theta_0) = |[\partial E\{S^*(\mathbf{X}, t)\}/\partial t]_{t=\theta} = \mu'(\theta_0, \theta_0)/\sigma(\theta_0, \theta_0) \qquad (1.6)$$

is, therefore a natural measure of its efficiency for testing H_0 against a close alternative H_1. If we put $\Delta = 1/\sqrt{n}$ then $e_S(\theta_0)$ is the displacement of the distribution of S from its null location under H_1, standardized with respect to its standard deviation under H_0.

Making the approximation $\sigma(\theta_0 + \Delta, \theta_0) \simeq \sigma(\theta_0, \theta_0)$ and assuming the relevant distributions of S to be approximately normal, the power of a level-α test of H_0 against H_1 is approximately

$$1 - \Phi\{u_\alpha - \Delta\sqrt{n}|\mu'(\theta_0, \theta_0)|/\sigma(\theta_0, \theta_0)\}, \tag{1.7}$$

where Φ is the standard normal distribution function.

For level-α tests based on S_1 and S_2 of H_0 against H_1 to have the same power, according to (1.7), the required sample sizes n_1 and n_2 must satisfy

$$n_1/n_2 = \{e_{S_2}(\theta_0)/e_{S_1}(\theta_0)\}^2 \tag{1.8}$$

The quantity $e_S(\theta_0)$ is called the *efficacy* of S at θ_0 and the ratio of squared efficacies in (1.8) is the *Pitman asymptotic relative efficiency* (ARE) of the statistics S_1 and S_2. In the 'regular' cases which we are considering the ARE is readily seen to be the same as the relative efficiency of the estimators yielded by S_1 and S_2. Further notes on the Pitman ARE are given in Appendix B.

1.7 Multiple samples and parameters

1.7.1 *Introduction*

The two-sample problem to receive most attention in later chapters is that of location shift. It is, therefore, still a one-parameter problem, and while the typical statistic used for inference will depend on two sets of sample values, the modifications required to the discussions in Sections 1.5 and 1.6 are obvious. They will be seen in the relevant later sections.

Problems with $k > 2$ samples are those that are commonly thought of as having to do with 'analysis of variance' in parametric statistics. Here a natural model involves $(k - 1)$ location-shift parameters, so we have a multiple-parameter problem. Regression generally involves multiple parameters, although the important straight-line regression case can be regarded as a one-parameter problem if the interest is only in the slope parameter. Two or more parameters also occur with bivariate or multivariate observations.

In the discussion that follows we shall consider two parameters. Generalization to more than two parameters is obvious. Also, as we have indicated for the two-sample, one-parameter case, there is no real need to discuss single- or multiple-sample cases separately at this stage; the actual details for these cases are, of course, different, but will become clearer when special cases are treated in later chapters.

Generally we shall have two statistics $S_r(\mathbf{X}, t_1, t_2)$, $r = 1, 2$ whose exact joint conditional distribution under a suitable randomization scheme is known when t_1 and t_2 are replaced by θ_1 and θ_2, the two parameters of interest. By 'known' we understand, as before, that the rule by which the distribution can be enumerated is known. In practice we may choose to approximate the distribution by, usually, a normal distribution. The statistics S_1 and S_2 will be assumed to have properties like those of S according to assumptions (i), (ii), (iii) in Section 1.6. Usually they will not be independent and we take their covariance to be of the form $\text{cov}\{S_1(\mathbf{X}, t_1, t_2),\ S_2(\mathbf{X}, t_1, t_2)\} = \sigma_{12}(t_1, t_2, \theta_1, \theta_2)/n$. As we have indicated in the introductory paragraph to this section, if more than one sample is involved, the only changes to the discussion are that the divisors, n, in the variances and covariances of S_1 and S_2 are replaced by factors depending on the sample sizes.

1.7.2 *Point estimation*

Defining the statistics S_1 and S_2 to be such that $E\{S_r(\mathbf{X}, \theta_1, \theta_2)\} = 0$, $r = 1, 2$ the point estimates of θ_1 and θ_2 are taken to be the solutions $\hat{\theta}_1, \hat{\theta}_2$ of the estimating equations

$$S_r(\mathbf{X}, t_1, t_2) = 0, \qquad r = 1, 2 \qquad (1.9)$$

If S_1 and S_2 are not continuous functions of t_1 and t_2 for fixed $\mathbf{X}$, some suitable convention has to be adopted to define unique estimates $\hat{\theta}_1, \hat{\theta}_2$.

The assumptions that have already been made about the distribution of S_1 and S_2 are sufficient to ensure the consistency of the estimates, by an argument similar to that leading to Lemma 1.1; it will not be elaborated here. Similarly, by arguments like those leading to (1.4) a formula for the large-sample approximate covariance matrix of $\hat{\theta}_1, \hat{\theta}_2$ can be obtained.

Let

$$\{\partial E\{S_r(\mathbf{X}, t_1, t_2)/\partial t_s\}_{t_1 = \theta_1, t_2 = \theta_2} = \gamma_{rs}(\theta_1, \theta_2)$$

and

$$C = \begin{bmatrix} \gamma_{11}(\theta_1, \theta_2) & \gamma_{12}(\theta_1, \theta_2) \\ \gamma_{21}(\theta_1, \theta_2) & \gamma_{22}(\theta_1, \theta_2) \end{bmatrix} \qquad (1.10)$$

Denote the covariance matrix of $S_1(\mathbf{X}, \theta_1, \theta_2), S_2(\mathbf{X}, \theta_1, \theta_2)$ by V. Then

$$\begin{bmatrix} \text{var}(\hat{\theta}_1) & \text{cov}(\hat{\theta}_1, \hat{\theta}_2) \\ \text{cov}(\hat{\theta}_1, \hat{\theta}_2) & \text{var}(\hat{\theta}_2) \end{bmatrix} \simeq C^{-1} V (C^{T})^{-1} \qquad (1.11)$$

By a linearization argument similar to that used before the joint distribution of $\hat{\theta}_1, \hat{\theta}_2$ is approximately normal if the joint distribution of S_1 and S_2 is approximately normal.

The efficiency of a point estimate will be judged in terms of its variance calculated according to (1.11).

1.7.3 Hypothesis testing

Let us consider the testing of a simple null hypothesis H that specifies the values (θ_1^0, θ_2^0) of θ_1 and θ_2, and the use of the statistics S_1 and S_2 for this purpose. Assuming that the exact joint null distribution, conditional or otherwise, of S_1 and S_2 is known from the operation of a suitable randomization scheme, the first matter that has to be decided is the choice of critical region for a test of level α. In the one-parameter case a suitable choice is usually obvious: $S(\mathbf{X}, t)$ is chosen not only to have zero expectation at $t = \theta$ but also such that $E\{S(\mathbf{X}, t)\}$ is monotonic in t. Then the appropriate tail region of the distribution of S defines a natural critical region. By extension of this sort of approach to the two-dimensional case, the critical region could be taken to comprise those points of the (S_1, S_2)-space that are 'most distant' from the origin $[E\{S(\mathbf{X}, \theta_1^0, \theta_2^0)\}, E\{S_2(\mathbf{X}, \theta_1^0, \theta_2^0)\}] = [0, 0]$.

Following precedents set in other areas of statistics, we shall use as a measure of distance

$$Q(X, \theta_1^\circ, \theta_2^\circ) = (S_1, S_2)V^{-1}(S_1, S_2)^{\mathrm{T}} \tag{1.12}$$

where we have written for simplicity of notation $S_r = S_r(\mathbf{X}, \theta_1^0, \theta_2^0)$, $r = 1, 2$, and V is the covariance matrix of S_1, S_2. Large values of Q will lead to rejection of H_0. In other words, we are now using Q as a test statistic for H_0.

Enumeration of the exact randomization distribution of Q is, in principle, straightforward and will be demonstrated in several examples to follow in later chapters. Therefore we have the apparatus for an exact test of H_0 based on S_1 and S_2.

Elementary calculations show that

$$E\{Q(\mathbf{X}, \theta_1^0, \theta_2^0)|H_0\} = 2$$

and this means that detailed enumeration of the distribution of Q will often be unnecessary; if the observed Q is smaller than 2, H_0 is accepted. Under suitable conditions the distribution of Q will be approximately χ_2^2.

The one-dimensional counterpart of Q is simply $S^2(\mathbf{X}, \theta_0)/\text{var } S(\mathbf{X}, \theta_0)$, whose expectation under H_0 is 1, and whose distribution under suitable conditions is approximately χ_1^2. Consistency of the test of H_0 can be shown to hold by arguing as in Section 1.3 but using the distribution of $S^2(\mathbf{X}, \theta_0)/\text{var}\{S(\mathbf{X}, \theta_0)\}$. A similar argument applied to Q demonstrates the consistency of the test of H_0 based on it.

For the two-dimensional case we define the efficacy of testing by generalization of the notion of small displacement of the distribution of S under H_0 relative to H_1 to two dimensions, using the type of distance measure that is involved in the definition of Q. By analogy with the interpretation of $e_S(\theta_0)$ in Section 1.6.2 as the displacement of the distribution of S due to a change of magnitude $1/\sqrt{n}$ in θ, we now perform a similar calculation. Both θ_1^0 and θ_2^0 are taken to be shifted by amounts $1/\sqrt{n}$ under H_1 and the displacement in $[E\{S_1(\mathbf{X}, \theta_1^0, \theta_2^0)\}, \; E\{S_2(\mathbf{X}, \theta_1^0, \theta_2^0)\}]$ is calculated according to the distance measure used to define Q. Then, if the efficacy of Q is $|e_Q(\theta_1^0, \theta_2^0)|$, we have

$$e_Q^2(\theta_1^0, \theta_2^0) = (1, 1)C^T V_*^{-1} C(1, 1)^T \tag{1.13}$$

where
$$V_* = \begin{bmatrix} \sigma_1^2(\theta_1^0, \theta_2^0) & \sigma_{12}(\theta_1^0, \theta_2^0) \\ \sigma_{12}(\theta_1^0, \theta_2^0) & \sigma_2^2(\theta_1^0, \theta_2^0) \end{bmatrix}$$

While developing an exact test procedure for a simple H_0 is hardly more complicated in the two-parameter case than for one parameter, severe complications can arise if we wish to test certain composite hypotheses. A typical example is the testing of a hypothesis that specifies, say, $\theta_1 = \theta_1^0$, but leaves θ_2 unspecified. The difficulty that arises is not peculiar to distribution-free tests. It is the general problem of existence of 'similar regions' or elimination of 'nuisance parameters'. In special cases it is possible to devise an exact test of such a null hypothesis despite the existence of the nuisance parameter θ_2. But in general it is not possible. An approximate procedure is to obtain a point estimate $\hat{\theta}_2$ of θ_2 and then to act as if $\hat{\theta}_2$ is the true value of θ_2. The accuracy of the resulting approximate test size is generally unknown.

1.7.4 Confidence regions

The argument for determining a confidence region outlined in Section 1.5 could be called inversion of the hypothesis-testing procedure. It carries over without modification to the two-parameter case, where inversion of the hypothesis-testing procedure for the simple H_0

produces a joint confidence region for the parameter vector (θ_1, θ_2). If the test of H_0 is exact, so is the confidence region.

Here, as in hypothesis testing, the problem of nuisance parameters arises, and again it is not confined to distribution-free methods. Let us suppose that an exact joint confidence region, C, for (θ_1, θ_2), of confidence coefficient $(1 - \alpha)$ has been established. Also assume for simplicity that it is convex. Suppose that this confidence region is graphed in a (θ_1, θ_2)-plane and that two tangents to it are drawn parallel to the θ_1 axis. They intersect the θ_2 axis at $\theta_2^{(1)}$ and $\theta_2^{(2)}$, and these two values can be taken as confidence limits for θ_2. However, the confidence coefficient of this confidence interval $(\theta_2^{(1)}, \theta_2^{(2)})$ for θ_2 is greater than or equal to $1 - \alpha$; usually it is greater than $1 - \alpha$ by an unknown amount. In 'normal theory' it is easy to establish the exact relationship between $1 - \alpha$ and the confidence coefficient of $(\theta_2^{(1)}, \theta_2^{(2)})$.

If two sets of tangents to C are drawn, parallel to the θ_2- and θ_1-axes respectively, confidence intervals for θ_1 and θ_2 are obtained. The probability that the two intervals simultaneously contain θ_1 and θ_2 is also *at least* $(1 - \alpha)$. Pairs of parallel tangents drawn at other angles, that is, not parallel to either of the axes, generate confidence intervals for linear functions of θ_1 and θ_2; the probability that all of these intervals simultaneously contain all of the relevant parameters is $1 - \alpha$. However interesting this statement may be, the most pressing practical problem is often that of obtaining a confidence interval for θ_1 or θ_2 or perhaps one linear function of θ_1 and θ_2, and determination of an exact interval appears to be impossible, in general, even with the availability of an exact joint confidence region C.

1.8 Normal approximations

1.8.1 *The need for normal approximations*

The inferential procedures described in this book are almost exclusively based on quite simple permutation and randomization schemes. Relevant conditional null distributions are obtainable exactly by these means, so that with patience or computing help they can always be listed in whatever detail is needed. So, from the point of view of doing distribution-free tests, or actually finding confidence limits, normal approximations are not strictly needed.

However, as some of the examples will show, the work of enumerating exact null distributions increases so rapidly with increasing sample sizes that putting distribution-free methods into effect would, in many instances, be quite impractical without the use of approximations. Knowledge of the first two moments of a null distribution will often be sufficient to enable one to make a sensible decision about a set of data. For example, in a test against a one-sided alternative the observed value of the test statistic may fall on the 'wrong' side of its null expectation. If the observed value of the test statistic is, say, more than four standard derivations from the null expectation there would be little question of accepting the null hypothesis.

Nevertheless, in most applications reasonably accurate significances are needed for tests, and they are, of course, needed for setting confidence limits. In calculations relating to these questions, good large-sample approximations to the exact distributions are almost indispensable. Another important aspect of normal approximations of distribution for statistics is that notions of asymptotic relative efficiency are closely connected with assumptions of normal distributions of statistics. It is fairly easy, and rather useful, to introduce crude ideas of efficiency without special reference to normal distribution, as we have done, but asymptotic normality certainly figures largely in more precise formulations.

In the following sections a brief outline of some of the important theory relevant to normal approximation of null distributions is given.

1.8.2 *The central limit theorem*

Suppose that the random variable has finite expectation μ and finite variance σ^2 and that $X_1, X_2, \ldots, X_n$ are independent random variables, identically distributed like X. Then we have the possibly best-known form of the central limit theorem as follows:

Theorem 1.3 If $Y_n = X_1 + X_2 + \ldots + X_n$, the distribution function of $(Y_n - n\mu)/(\sigma\sqrt{n})$ converges to a standard normal ($N(0, 1)$) distribution function as $n \to \infty$.

Theorem 1.3 is the basis of the normal approximations of the distributions of sums of independent random variables that are used widely in practice. Even when the actual distribution is discrete the approximation can be remarkably good for quite small values of n, especially if appropriate continuity corrections are made. One of the best-known examples, which we shall be using, is normal approximation of the Binomial distribution.

A more general version of Theorem 1.3 holds for independent random variables that are not necessarily identically distributed. We shall state it in a form given and proved in Hájek and Sidák (1967). Suppose that $a_1, a_2, \ldots, a_n$ is a sequence of real numbers with the property

$$\max_i (a_i^2) / \sum_i a_i^2 \to 0 \qquad \text{as} \quad n \to \infty \qquad (1.14)$$

and that $X_1, X_2, \ldots, X_n$ are the random variables of Theorem 1.3. Let

$$T_a = \sum_{i=1}^{n} a_i X_i$$

Then for any finite n,

$$\mu_a = E(T_a) = \mu \sum_{i=1}^{n} a_i; \qquad \sigma_a^2 = \text{var}(T_a) = \sigma^2 \sum_{i=1}^{n} a_i^2.$$

Theorem 1.4 The distribution function of $(T_a - \mu_a)/\sigma_a$ converges to an $N(0, 1)$ distribution function as $n \to \infty$.

There will be several applications of this theorem in later chapters. One of them is to the distribution of the Wilcoxon signed-rank statistic.

1.8.3 *Sampling from finite populations*

In problems involving two samples of sizes m and n it is common to test a null hypothesis, H_0, to the effect that the two samples derive from the same population. Some cases are exactly of this kind, many others can be put in this form. Here we obtain exact inferential procedures by conditioning on the observed set of $N = m + n$ results, and arguing that, under H_0 the partition into groups of sizes m and n is random. This argument is used throughout Chapter 4 and elsewhere.

The randomization procedure here can be seen to be equivalent to sampling without replacement from a finite population of size $N = m + n$, the sample size being m or n. Suppose that the N sample items in such a finite population have values $\xi_1, \xi_2, \ldots, \xi_N$, with

$$\mu_N = (1/N) \sum_{i=1}^{N} \xi_i, \qquad \sigma_N^2 = (1/N) \sum_{i=1}^{N} (\xi_i - \mu_N)^2 \qquad (1.15)$$

Now let $X_1, X_2, \ldots X_m$ be the values of the members of a sample of size m drawn at random without replacement from the population. By symmetry, the random variables $X_1, X_2, \ldots, X_m$ are identically distributed, each with mean μ_N and variance σ_N^2. Nothing that var $(X_1 + \ldots + X_N) = 0$ we find that cov $(X_i, X_j) = -\sigma_N^2/(N-1)$, all i, j and hence the following standard results are readily established:

$$E\left(\sum_{i=1}^{m} X_i\right) = m\mu_N, \qquad \text{var}\left(\sum_{i=1}^{m} X_i\right) = m\sigma_N^2(N-m)/(N-1)$$

they are used repeatedly, especially in Chapter 4.

A general theorem of Wald and Wolfowitz (1944) deals with the question of asymptotic normality of linear functions of $X_1, X_2, \ldots, X_N$. A special case is the sampling problem discussed above, and we have the following theorem given also in Wilks (1962, p. 268).

Theorem 1.5 Suppose that we have a sequence of finite populations Π_N and that for large N

$$(1/N) \sum_{i=1}^{N} (\xi_i - \mu_N)^r/\sigma_N^r = O(1) \tag{1.16}$$

Then if $m, N \to \infty$ in such a way that $N/m \to c$, where $1 < c < \infty$, the distribution function of

$$\left(\sum_{i=1}^{m} X_i - m\mu_N\right) \bigg/ \{\sigma_N^2 mn/(N-1)\}^{1/2}$$

converges to a standard normal distribution function.

The more general version of this theorem deals with linear functions of the type

$$L_N = \sum_{j=1}^{N} a_j X_j$$

where $(X_1, X_2, \ldots, X_N)$ is uniformly distributed on all permutations of the values $(\xi_1, \xi_2, \ldots, \xi_N)$. This theorem is widely applicable, and could be used to check the asymptotic normality of certain rank statistics which we shall use.

For the $(a_1, a_2, \ldots, a_N)$ and $(\xi_1, \xi_2, \ldots, \xi_N)$ sequences we introduce parameters $\mu_N(a), \sigma_N^2(a), \mu_N(\xi), \sigma_N^2(\xi)$, their respective means and variances, defined according to (1.15). Then using the result given above for $\text{cov}(X_i, X_j)$ it is easily shown that

$$E(L_N) = N\mu_N(\xi)\mu_N(a)$$

$$\text{var}(L_N) = \left(\frac{N}{N-1}\right)N\sigma_N^2(\xi)\sigma_N^2(a)$$

Theorem 1.6 Suppose that conditions like (1.16) hold for each of the sequences $(a_1, a_2, \ldots, a_N)$ and $(\xi_1, \xi_2, \ldots, \xi_N)$, and that $m, N \to \infty$ such that $N/m \to c$, $1 < c < \infty$. Then the distribution function of $\{L_N - E(L_N)\}/\{\text{var}(L_N)\}^{1/2}$ converges to a standard normal distribution function.

This theorem, with an outline of the proof, is given in Wilks (1962, p. 266).

A bivariate extension of this theorem can be developed if we take each member of the population to have two measurements (ε_i, η_i), $i = 1, 2, \ldots, N$. Then write $L_{Na} = \sum a_j X_j$ and $L_{Nb} = \sum b_j Y_j$, with Y_j defined similarly for X_j.

By imposing suitable conditions on the four relevant sequences, applying Theorem 1.6 to linear functions of L_{Na} and L_{Nb}, and using a characterization of the bivariate normal distribution the joint asymptotic normality of L_{Na} and L_{Nb} can be established.

1.8.4 *Linear rank statistics*

Suppose that $X_1, X_2, \ldots, X_n$ are independent and identically distributed with the common distribution function F, and denote the rank of X_i among the values $X_1, X_2, \ldots, X_n$ by $R_i, i = 1, 2, \ldots, n$. Many of the statistics that we shall discuss are of the form

$$W = \sum_{i=1}^{n} c_i R_i$$

The null distributions of these statistics are usually based on the randomization scheme whereby all permutations of the ranks $(1, 2, \ldots, n)$ have equal probability. Then, by the methods used in Section 1.8.3, we have the following familiar results:

$$E(R_i) = (n+1)/2, \qquad i = 1, 2, \ldots, n$$
$$\text{var}(R_i) = \sigma^2 = (n^2 - 1)/12, \qquad i = 1, 2, \ldots, n$$
$$\text{cov}(R_i, R_j) = -\sigma^2/(n-1), \qquad i \neq j = 1, 2, \ldots, n$$

Using these formulae we obtain

$$E(W) = \left(\frac{n+1}{2}\right) \sum_{i=1}^{n} c_i$$

$$\text{var}(W) = \sigma^2 \left(\frac{n}{n-1}\right) \sum_{i=1}^{n} (c_i - \bar{c})^2 = \frac{n(n+1)}{12} \sum_{i=1}^{n} (c_1 - \bar{c})^2$$

where $\bar{c} = (1/n) \sum_{i=1}^{n} c_i$.

The following theorem is a special case of a theorem for rank-based scores given in Hájek and Sidák (1967), Ch. V:

Theorem 1.7 If $\max_i (c_i - \bar{c})^2 / \sum_{i=1}^{n} (c_i - \bar{c})^2 \to 0$ as $n \to \infty$, the distribution function of $(W - E(W))/\{\text{var}(W)\}^{1/2}$ converges to a standard normal distribution function.

CHAPTER 2

One-sample location problems

2.1 Introduction

Throughout this chapter we shall suppose that n independent observations $x_1, x_2, \ldots, x_n$ are made on a random variable X whose cumulative distribution function (c.d.f.) is $F(x)$, where $F(x)$ is continuous with differentiable probability density function (p.d.f.) $f(x)$. Our concern will be primarily with the location of F, and when appropriate, we shall write $F(x) = F(x, \theta)$, where θ is a suitably defined location parameter. The simpler notation, $F(x)$, will be used unless it is essential to indicate parameter values explicitly.

We shall consider questions of point and interval estimation of θ, and of testing hypotheses about θ. Estimating equations for θ will be formulated, and the approach to testing of hypotheses, and finding confidence limits, will be to use the statistics appearing in these equations.

The best-known measures of location of a distribution F are the mean and the median, each being a natural measure in its own way. In the distribution-free setting the mean is less suitable than the median because, without some restriction on the family of distributions to which F may belong, the mean may not exist. On the other hand, a median always exists. Consequently, distribution-free inferences about location are most commonly thought of as dealing with the median. Unless otherwise specified, we shall take θ to be the population median in the remainder of this chapter.

If the distribution F is symmetrical about a point θ such that $f(x - \theta) = f(-x + \theta)$ then θ will also be referred to as the 'centre' of the distribution; if F is continuous, the mean, if it exists, and the median, suitability defined, coincide with θ. Some of the important one-sample distribution-free techniques depend on symmetry of F, and are, therefore, distribution-free in a restricted sense. The class of problems involving symmetric F is, however, quite extensive. Instances of naturally occurring symmetric distributions are not rare, and symmetric distributions are often produced by differencing of independent random variables, for example, as in 'paired comparison' trials.

The mean, μ, is defined by

$$\mu = \int x \, dF(x) \tag{2.1}$$

when the r.h.s. of (2.1) is finite. With suitable $F(x)$ it is sometimes convenient to define it as the value of t that minimizes

$$Q_2(t) = \int (x - t)^2 \, dF(x) \tag{2.2}$$

The median θ satisfies $F(\theta) = \frac{1}{2}$ and, with suitable $F(x)$, can also be defined as the value of t that minimizes

$$Q_1(t) = \int |x - t| \, dF(x) \tag{2.3}$$

A direct way of using the observations to estimate μ or θ is to replace the distribution function in (2.1), (2.2) or (2.3) by the sample c.d.f. From (2.1) we then find by simple calculations that the estimate of μ is

$$\hat{\mu} = \frac{1}{n} \sum_{i=1}^{n} x_i = \bar{x}$$

The sample version of (2.3) can be written

$$Q_{1n}(t) = \frac{1}{n} \sum_{i=1}^{n} |x_{(i)} - t| \tag{2.4}$$

where $x_{(1)} < x_{(2)} < \ldots < x_{(n)}$ are the ordered $x_1, x_2, \ldots, x_n$. When graphed as a function of $t, Q_{1n}(t)$ has slope

$$\frac{1}{n} S(\mathbf{x}, t) = \frac{1}{n} \sum_{i=1}^{n} \text{sgn}\,[x_{(i)} - t] \tag{2.5}$$

for values of t not coinciding with the order statistics $x_{(1)}, x_{(2)}, \ldots, x_{(n)}$. This is a step function with steps of equal height $2/n$ occurring at $x_{(1)}, x_{(2)}, \ldots, x_{(n)}$.

The minimum of $Q_{1n}(t)$ occurs at a value $t = \hat{\theta}$ where $S(\mathbf{x}, t)$ changes sign. When $n = 2k + 1$ we have $\hat{\theta} = x_{(k+1)}$. When $n = 2k$, $S(\mathbf{x}, t) = 0$ for $x_{(k)} < t < x_{(k+1)}$ and the usual convention is to put $\hat{\theta} = \{x_{(k)} + x_{(k+1)}\}/2$. Formally we may regard $\hat{\theta}$ as the solution of

$$S(\mathbf{x}, t) = 0$$

Since $F_n(t) = \left\{ -\dfrac{1}{n}S(\mathbf{x}, t) + 1 \right\} \Big/ 2$ is the sample c.d.f., $\hat{\theta}$, also satisfies $F_n(\hat{\theta}) = \frac{1}{2}$.

Other measures of location can be defined by different choices of Q, the function to be minimized. Alternatively, different statistics could be used in (2.5). We might, for example, define a location estimate as the value of t for which

$$S_\psi(\mathbf{x}, t) = \frac{1}{n} \sum_{i=1}^{n} \psi(x_i - t) = 0 \qquad (2.6)$$

is satisfied. If it happens that $\psi(u) = f'(u)/f(u)$, and $f(x - \theta)$ is the p.d.f. of X, then the maximum-likelihood (ML) estimate (MLE) of θ is obtained. For this reason, estimates of the type generated by (2.6) are called M-estimates (Huber (1972)). The equation in t,

$$E\psi(\mathbf{X}, t) = \int \psi(x - t)\, \mathrm{d}F(x) = 0 \qquad (2.7)$$

can be taken to define a location parameter of F.

2.2 The median

The median was defined in Section 2.1 and, in the terminology of Chapter 1, the estimating equation for the median, motivated directly by the definition, is (2.5), which we rewrite as

$$S(\mathbf{x}, t) = \sum_{i=1}^{n} \operatorname{sgn}(x_i - t) = 0 \qquad (2.8)$$

In discussions of the median we shall take the population median to be θ unless otherwise stated, and in this case,

$$\Pr\left[\operatorname{sgn}(X_i - \theta) = +1\right] = \Pr\left[\operatorname{sgn}(X_i - \theta) = -1\right] = \tfrac{1}{2},$$
$$i = 1, 2, \ldots, n \qquad (2.9)$$

giving

$$E\{S(\mathbf{X}, \theta)\} = 0 \qquad (2.10)$$

Equation (2.10), by an application of the method of moments, can be taken as an alternative motivation for the estimating equation (2.8).

The null distribution of S

Here and elsewhere the term 'null distribution' indicates the distribution of a statistic $S(\mathbf{X}, \theta)$ when the distribution of X has parameter value θ. Using (2.9) and the independence of the random variables

$\mathrm{sgn}\,(X_i - \theta)$, $i = 1, 2, \ldots n$, it is easy to enumerate the null distribution of S. Since we can write $S(\mathbf{X}, \theta) = 2(B - n/2)$ where B is a binomial $(n, \tfrac{1}{2})$ random variable, tables of the binomial distribution can be used to find the null distribution of S. Note that this distribution of $S(\mathbf{X}, \theta)$ does not depend on the form of F, hence inference based on S is distribution-free.

Since $S(\mathbf{X}, \theta)$ is the sum of independent random variables identically and symmetrically distributed about 0, its distribution is symmetric about 0, and since $\mathrm{var}\,\{\mathrm{sgn}\,(X - \theta)\} = 1$,

$$\mathrm{var}\,\{S(\mathbf{X}, \theta)\} = n \qquad (2.11)$$

By an application of the central limit theorem, the distribution of $S(\mathbf{X}, \theta)$ tends to normality as n increases.

Testing $H_0 : \theta = \theta_0$

To test $H_0 : \theta = \theta_0$ against $H_1 : \theta = \theta_1 > \theta_0$, we calculate the observed value $S(\mathbf{x}, \theta_0)$ of $S(\mathbf{X}, \theta_0)$ and refer it to the null distribution of S. Since $\mathrm{Pr}\,\{\mathrm{sgn}\,(X - \theta_0) = +1 | H_1\} > \tfrac{1}{2}$, we have $E\{S(X, \theta_0) | H_1\} > 0$, and therefore we reject H_0 if $S(\mathbf{x}, \theta_0)$ is in the upper tail of the null distribution. In practice one might refer to a table of the null distribution, or for $n \geqslant 5$, use a normal approximation as follows:

Suppose $S(\mathbf{x}, \theta) = s$. Calculate

$$u = \frac{s - 1}{\sqrt{n}} \qquad (2.12)$$

and reject H_0 at level α if $1 - \Phi(u) < \alpha$. In (2.12), the subtraction of 1 from s in the r.h.s. numerator is a continuity correction, because the c.d.f. of S is discrete with jumps at $-n, -n+2, \ldots, n-2, n$.

Since $E\{S(\mathbf{X}, \theta_0) | H_1\} = 2n\{F(\theta_1, \theta_1) - F(\theta_0, \theta_1)\}$ and var $\{S(\mathbf{X}, \theta_0) | H_1\} \leqslant n$, the test is consistent by an application of the results of Section 1.3.

For other H_1, modifications of the test procedure are obvious.

Example 2.1 Suppose that we have the $n = 10$ independent observations.

$$3.38,\ 5.81,\ 4.46,\ 4.62,\ 4.15,\ 5.44,\ 6.56,\ 5.82,\ 3.95,\ 5.19$$

and wish to test the hypothesis $H_0 : \theta = 4.25$ against $H_1 : \theta > 4.25$, where θ is the population median. Then, $S(\mathbf{x}, 4.25) = 4$, and from tables, or by simple enumeration, $\mathrm{Pr}[S(\mathbf{X}, \theta_0) \geqslant +4] = 176/1024$.

The value of u given by (2.12), is $3/\sqrt{10} = 0.95$ giving $1 - \Phi(u) = 0.171$, which agrees well with the exact significance level $176/1024 = 0.172$.

Confidence limits for θ

A simple way of finding a confidence set for θ is to use the following procedure of inverting the hypothesis-testing argument: A $100(1 - \alpha)\%$ confidence set, C, for θ comprises those values θ' for which the hypothesis that $\theta = \theta'$ is accepted at level α. Typically the form of critical region in the hypothesis test determines the form of C, in particular, whether one- or two-sided confidence limits are obtained.

For two-sided confidence limits we may proceed as follows. Let $\Pr[|S_1(\mathbf{X}, \theta)| \leqslant s] = 1 - \alpha$, where α is near one of the conventional 'small' values 0.10, 0.05, etc. Then, as t varies from $-\infty$ to $+\infty$, observed $S(\mathbf{x}, t)$ changes from n to $-n$, in jumps of size 2, at points $x_{(1)}, x_{(2)}, \ldots, x_{(n)}$. Consequently, the limits are $x_{(r_1)} +$ and $x_{(r_2)} -$ where $r_1 = (n - s)/2$, $r_2 = n - r_1 + 1$.

For large n, the values s, r_1, r_2 can be found by the normal approximation for the distribution of S.

Example 2.2 Use the data in Example 2.1 arranged in increasing order of magnitude. The bracketed numbers separating the observed order statistics are values of S given by t in the corresponding interval

(10) 3.38   (8)   3.95   (6)   4.15   (4)   4.46   (2)   4.62   (0)

(0)   5.19   (−2)   5.44   (−4)   5.81   (−6)   5.82   (−8)   6.56   (−10)

From the null distribution of S we have

$$\Pr[|S| \leqslant 4] = -112/1024 \simeq 0.90.$$

Therefore a $100(1 - 112/1024)\%$ confidence interval for θ is

$$(4.15 +, 5.81 -)$$

Point estimation of θ

The point estimate $\hat{\theta}$ of θ is the solution of the estimating equation (2.8), and, as explained before we have

$$\hat{\theta} = \begin{cases} x_{(k+1)} & \text{if } n = 2k + 1 \\ \{x_{(k)} + x_{(k+1)}\}/2 & \text{if } n = 2k \end{cases}$$

the symbol $\hat{\theta}$ will be used to denote the estimate and the corresponding estimator.

To check for consistency of $\hat{\theta}$ as an estimate of θ we can use the result in Section 1.4. We only have to verify that var$\{S(\mathbf{X}, t)\} = n\sigma^2(t, \theta)$ with $\sigma^2(t, \theta)$ bounded. In fact

$$\text{var}\{S(\mathbf{X}, t)\} = 4n\,\text{Pr}\,\{\text{sgn}\,(X - t) = +1\} \cdot \text{Pr}\,\{\text{sgn}\,(X - t) = -1\} \leqslant n$$

therefore $\hat{\theta}$ is consistent for θ.

In general, the estimator $\hat{\theta}$ is biased for θ, but it is median-unbiased, that is, the median of $\hat{\theta}$ is θ.

For the case $n = 2k + 1$, the p.d.f. of $\hat{\theta}$ at $\theta = t$ is

$$\{n!/(k!)^2\}\,\{F(t)(1 - F(t))\}^k$$

and by using Taylor series approximations of the type $F(t) = F(\theta) + (t - \theta)f(\theta) = \frac{1}{2} + (t - \theta)f(\theta)$ it is possible to show that the p.d.f. of $\hat{\theta}$ can be approximated by an $N(\theta, 1/(4nf^2(\theta)))$ density as $n \to \infty$. See, for example, David (1970, p. 201).

Alternatively, we may use the results in Section 1.6.1 as follows. We have

$$E\{S(\mathbf{X}, t)\} = n\{1 - 2F(t)\}$$

giving

$$\left(\frac{\partial E\{S(\mathbf{X}, t)\}}{\partial t}\right)_{t=\theta} = -2nf(\theta) \tag{2.13}$$

and hence

$$\text{var}(\hat{\theta}) \simeq \text{var}\{S(\mathbf{X}, \theta)\}/\{\partial E\{S(\mathbf{X}, t)\}/\partial t\}^2_{t=\theta} = 1/\{4nf^2(\theta)\} \tag{2.14}$$

Since the distribution of $S(\mathbf{X}, \theta)$ is approximately normal for large n, the distribution of $\hat{\theta}$ is also approximately normal for large n.

Efficiency considerations
We begin by considering efficiency of hypothesis testing using S. Recalling the definition of efficacy given in Section 1.6.2 the efficacy of S is

$$e_S(\theta) = |\{\partial E\{S(\mathbf{X}, t)\}/\partial t\}_{t=\theta}|/\{n\,\text{var}\,S(\mathbf{X}, \theta)\}^{1/2}$$

and using (2.11) and (2.13) we obtain

$$e_S(\theta) = 2f(\theta) \tag{2.15}$$

In order to get a quantitative impression of the efficiency of S and its associated point estimate, relative to other statistics, particular examples of F have to be considered. In the examples that follow, S is compared with the appropriate likelihood procedure. With twice-differentiable likelihoods this procedure yields the estimating equation

$$\sum_{i=1}^{n} \frac{1}{f(x_i, t)} \frac{\partial f(x_i, t)}{\partial t} = 0 \qquad (2.16)$$

and its efficacy is

$$e_{ML}(\theta) = \left| \left(E \frac{\partial^2 \ln f(\mathbf{X}, t)}{\partial t^2} \right)_{t=\theta} \right|^{1/2} \qquad (2.17)$$

Example 2.3 If the distribution of X is $N(\theta, \sigma^2)$, we have

$$e_S(\theta) = (2/\pi)^{1/2}/\sigma$$

$$e_{ML}(\theta) = 1/\sigma$$

giving Pitman ARE $= 2/\pi$.

Example 2.4 $F(x, \theta) = 1 - e^{-x \ln 2/\theta}, x \geqslant 0$; this is an exponential distribution with median θ and mean $\theta/\ln 2$.

$$e_S(\theta) = \left(\frac{\ln 2}{\theta} \right)$$

$$e_{ML}(\theta) = \frac{1}{\theta}$$

$$\text{ARE} = (\ln 2)^2$$

Since the sample mean is a commonly used location estimator, a comparison of its efficiency relative to the median is of interest. In Example 2.3 the MLE is the sample mean and in Example 2.4 the MLE of θ is (sample mean/ln 2). Consequently the ARE results reported in these examples also apply to the sample mean. One might expect the sample median to be relatively more efficient than the sample mean for 'heavy-tailed' distributions, because the sample median is little influenced by fluctuations in the smallest and largest sample values. Two examples follow.

Example 2.5 $f(x, \theta) = (1/\pi)\{1 + (x - \theta)^2\}$; a Cauchy distribution. This is a somewhat extreme example since the mean is undefined and $\text{var}(\bar{X})$ is infinitely large.

$$e_S(\theta) = 2/\pi = 0.64; \qquad e_{ML}(\theta) = 0.71; \qquad e_{\bar{X}}(\theta) = 0$$

Thus, relative to the MLE, the sample median has the moderately high efficiency of 81%, whereas the sample mean is totally inefficient.

Example 2.6 $f(x, \theta) = \dfrac{18}{20} \phi(x - \theta) + \dfrac{2}{20} \phi\left(\dfrac{x - \theta}{10}\right)$ where $\phi(u)$ is the standard normal p.d.f.; this is one of the examples used by Huber (1972) in robustness studies. Efficacies are

$$e_S(\theta) = 0.73; \qquad e_{ML}(\theta) = 0.91; \qquad e_{\bar{x}}(\theta) = 0.30$$

In this less extreme example the median is again considerably more efficient than the mean.

In concluding this section we emphasize that the efficiency of the median is strongly dependent on the behaviour of $f(x, \theta)$ for x near θ, and it is easy to construct examples for which the median would be disastrously bad or, again, extremely good relative to the mean. Overall, $S(\mathbf{X}, t)$ should be a reasonably robust statistic within the class of unimodal distributions.

Regarding efficiency of estimation, we have already seen that the large-sample variance of the sample median is given by equation (2.14). The large-sample variance of the MLE of θ is $\{e_{ML}(\theta)\}^{-2}/n$ when (2.17) holds. Thus we see that if we measure relative efficiency of estimation by the ratio of large-sample variances, then it is equal to the Pitman ARE of testing.

2.3 Symmetric distributions

2.3.1 *A basic permutation argument*

If F is known to be symmetric, several distribution-free procedures, apart from those discussed in Section 2.2, become available. Of course, these procedures are distribution-free only within the class of symmetric distribution F. They depend on the following permutation argument used by Fisher (1966, p. 41).

With the distribution of X symmetric about θ, given an observation x such that the magnitude of its distance from θ is $|x - \theta|$, the sign of the difference is positive or negative with equal probability, $\frac{1}{2}$. Therefore, if we put

$$Y = \text{sgn}(x - \theta)|x - \theta| \qquad (2.18)$$

then, conditionally on $|X - \theta| = |x - \theta|$, Y has a two-point distribution with

$$\Pr[Y = -|x - \theta|] = \Pr[Y = +|x - \theta|] = \tfrac{1}{2} \qquad (2.19)$$

Suppose, now, that we have n independent observations $x_1, x_2, \ldots, x_n$ and define $Y_1, Y_2, \ldots Y_n$ according to (2.18). Then the joint distribution of $Y_1, Y_2, \ldots, Y_n$, conditional on the observed $|x_i - \theta|$, $i = 1, 2, \ldots, n$, is easily enumerated by listing the 2^n sign arrangements of the type $(+ + - \ldots -)$. Consequently it is also easy, although tedious for large n, to find the exact conditional distribution of any function of the Y_i that may be considered for use in inference about θ.

2.3.2 The mean statistic $A(\mathbf{X}, \theta)$

Let

$$A(\mathbf{x}, \theta) = \sum_{i=1}^{n} Y_i = \sum_{i=1}^{n} \operatorname{sgn}(x_i - \theta)|x_i - \theta| \qquad (2.20)$$

This $A(\mathbf{x}, \theta)$ is called the *mean statistic* because we could write it simply

$$A(\mathbf{x}, \theta) = \sum_{i=1}^{n} (x_i - \theta)$$

whence the corresponding estimating equation is readily seen to have as its solution the sample mean $\bar{x}$. Inference about θ using A is, therefore, tantamount to using the sample mean. However, since we shall make use of the conditional distribution of A, obtained by the permutational method described in Section 2.3.1, the inferences are exactly conditionally distribution-free. Significance levels are conditionally exact, and since the unconditional significance levels are just expectations of the conditional levels they are also exact.

Testing H_0 using A
In testing $H_0 : \theta = \theta_0$ against $H_1 : \theta > \theta_0$, we note that unconditionally

$$E\{A(\mathbf{X}, \theta_0)|\theta = \theta_0\} = 0$$

and

$$E\{A(\mathbf{X}, t)|\theta = \theta_0\} > 0, \quad \text{if} \quad t > \theta_0$$

if these expectations exist. Thus the critical region for testing H_0 against H_1 is chosen so as to reject H_0 if the observed $A(\mathbf{x}, \theta_0)$ is sufficiently large.

Example 2.7 We have $n = 6$ observations:

$$3.38, 5.81, 4.46, 4.62, 4.15, 5.44$$

Assume that they originate from a symmetric distribution and let $\theta_0 = 4.5$. We shall test at level $6/64$ (approximately 10%), and hence need only list the 6 largest values of $A(\mathbf{x}, \theta_0)$ as follows:

$\|x_i - \theta_0\|$:	1.12	1.31	0.04	0.12	0.35	0.94	
	+	+	+	+	+	+	$A = 3.88$
	+	+	−	+	+	+	$A = 3.80$
	+	+	+	−	+	+	$A = 3.64$
	+	+	−	−	+	+	$A = 3.56$
	+	+	+	+	−	+	$A = 3.18$
	+	+	−	+	−	+	$A = 3.10$
observed:	−	+	−	+	−	+	$A = 0.86$

Since observed $A = 0.86$ is smaller than the six largest values of A, we accept H_0 at level $6/64$.

Confidence limits using A

Let $N(t)$ denote the number of possible $A(\mathbf{X}, t)$ values greater than the observed $A(\mathbf{x}, t)$. The possible $A(\mathbf{X}, t)$ values are generated by the 2^n permutations as described in Section 2.3.1; we denote this set of permutations by Q, while Q' denotes the $2^n - 1$ permutations obtained by removing the observed one from Q. A $100(1 - 2r/2^n)\%$ two-sided confidence interval for θ is (t_1, t_2), with t_1 determined such that for $t \lessgtr t_1$ we have $N(t) \lessgtr r$; t_2 is determined similarly.

As t varies from $-\infty$ to $+\infty$, $N(t)$ changes whenever t passes through one of all the possible averages $(x'_1 + x'_2 + \ldots + x'_s)/s$, where $1 \leqslant s \leqslant n$ and $x'_1, x'_2, \ldots, x'_s$ is a subset of the sample values $x_1, x_2, \ldots, x_n$. This can be seen by writing

$$N(t) = \sum_{q \in Q'} I\left[\sum_q \{\operatorname{sgn}(x_i - t) - \operatorname{sgn}(X_i - t)\} |x_i - t| < 0 \right]$$

where q denotes a member of Q', and $I(\ )$ is the indicator function. Now

$$\{\operatorname{sgn}(x_i - t) - \operatorname{sgn}(X_i - t)\} |x_i - t|$$
$$= \begin{cases} 0 & \text{if } \operatorname{sgn}(x_i - t) = \operatorname{sgn}(X_i - t) \\ 2(X_i - \theta) & \text{if } \operatorname{sgn}(x_i - t) \neq \operatorname{sgn}(X_i - t) \end{cases}$$

Therefore

$$N(t) = \sum_{Q'} I\left(\frac{x'_1 + x'_2 + \ldots + x'_s}{s} < t \right)$$

showing that $N(t)$ changes at the values $(x'_1 + \ldots + x'_s)/s$.

Example 2.8 We use the data of Example 2.7. For an approximately

90% two-sided confidence interval, we have $r = 3$. The three largest averages are

$$5.81$$
$$(5.81 + 5.44)/2 = 5.625$$
$$5.44$$

The three smallest averages are

$$3.38$$
$$(3.38 + 4.15)/2 = 3.765$$
$$(3.38 + 4.46)/2 = 3.92$$

Thus the $100(1 - 6/64)\%$ confidence interval for θ is: $(3.92, 5.44)$.

Normal approximation of A

For the calculation of significance levels in large samples it would be useful to be able to approximate the conditional distribution of A by a normal distribution. Since conditionally A is the sum of independent, not identically distributed random variables, such an approximation is possible under certain conditions; see Section 1.8. The variance of the conditional distribution of A is $\sum_{i=1}^{n}(x_i - t)^2$ and approximation by a $N(0, \sum_{i=1}^{n}(x_i - t)^2)$ distribution will be possible, almost surely, if $\max_i (X_i - t)^2 / \sum_{i=1}^{n}(X_i - t)^2 \to 0$, almost surely, as $n \to \infty$. This condition will hold for distributions F with finite variance, but not, for example, in the case of the Cauchy distribution.

To illustrate the use of the normal approximations, consider the one-sided test discussed above with significance level α. We reject H_0 if

$$\sum_{i=1}^{n}(x_i - \theta_0) > u_\alpha \left[\sum_{i=1}^{n}(x_i - \theta_0)^2 \right]^{1/2} \tag{2.21}$$

where $\Phi(u_\alpha) = 1 - \alpha$. Relation (2.21) can be rewritten

$$\bar{x} > \theta_0 + \frac{s}{\sqrt{n}} u_\alpha \left(\frac{n-1}{n - u_\alpha^2} \right)^{1/2}$$

where $s^2 = \sum_{i=1}^{n}(x_i - \bar{x})^2 / (n-1)$.

The factor $u_\alpha \{(n-1)/(n - u_\alpha^2)\}^{1/2}$ can be taken as an approximation to $t_{n-1}(1 - \alpha)$, which would have been used in its place if the observations were known to have originated from an $N(\theta, \sigma^2)$ population. For certain n and α values the approximation is quite good; for example, with $\alpha = 0.05$ we have the following:

n	$t_{n-1}(0.95)$	$1.645\{(n-1)/(n-1.645^2)\}^{1/2}$
5	2.132	2.172
10	1.833	1.827
30	1.699	1.696

Point estimation based on A and efficiency considerations

In the introduction to Section 2.3.2 we mentioned that the point estimate of θ resulting from the estimating equation defined by A is the sample mean $\bar{x}$. While it is possible to make exact conditional distribution-free inferences, unconditional behaviour must be studied when assessing the efficiency of a procedure based on A relative to other statistics.

The efficacy of A is

$$e_A(\theta) = 1/\sigma$$

where $\sigma^2 = \text{var}(X)$ if it exists; and, $\text{var}(\bar{X}) = \sigma^2/n$.

2.3.3 *Rank transformations*

Comparing the statistics S and A discussed in previous sections, it will be noted that S can be regarded as having been derived from A by applying a transformation $T(u) = 1$ to the values of $|x_i - \theta|$ in A. Thus

$$S(\mathbf{X}, \theta) = \sum_{i=1}^{n} \text{sgn}(X_i - \theta)T(|X_i - \theta|) \qquad (2.22)$$

Many other transformations are possible; one of the best known, and most effective, from the point of view of efficiency and robustness, is the rank transformation where $|x_i - \theta|$ is replaced by its rank, denoted $\text{Rank}(|x_i - \theta|)$, in the set $|x_1 - \theta|, |x_2 - \theta|, \ldots, |x_n - \theta|$. Recall that in a set of numbers $a_1, a_2, \ldots, a_n$, $\text{Rank}(a_i) = \#(a_j \leqslant a_i, j = 1, 2, \ldots, n)$. The statistic produced by this rank transformation is

$$W(\mathbf{X}, \theta) = \sum_{i=1}^{n} \text{sgn}(X_i - \theta)\,\text{Rank}(|X_i - \theta|) \qquad (2.23)$$

called the *Wilcoxon signed-rank statistic*, after Wilcoxon (1945).

The null distribution of W

Owing to the symmetry of the distribution of X about θ, the random variables $|X_i - \theta|$ and $\text{sgn}(X_i - \theta)$ are independent; this holds because $\Pr[\text{sgn}(X_i - \theta) = -1] = \Pr[\text{sgn}(X_i - \theta) = +1] = \frac{1}{2}$ for any given $|X_i - \theta|$. Writing $S_i(\theta) = \text{sgn}(X_i - \theta)$, $R_i(\theta) = \text{Rank}(|X_i - \theta|)$, it follows that $S_1(\theta), S_2(\theta), \ldots, S_n(\theta)$ are mutually independent, and $S_i(\theta)$ is independent of $R_j(\theta)$ for $i, j = 1, 2, \ldots, n$; note, however, that $R_1(\theta), R_2(\theta), \ldots, R_n(\theta)$ are not mutually independent.

In the joint distribution of $R_1(\theta), R_2(\theta), \ldots, R_n(\theta)$, the vector $\mathbf{R}(\theta) = (R_1(\theta), R_2(\theta), \ldots, R_n(\theta))$ assumes each of the $n!$ permutations of the integers $1, 2, \ldots, n$ with equal probability. Using this fact and the independence properties noted above, the distribution of W can be enumerated by listing the 2^n possible values of the vector $(S_1(\theta), S_2(\theta), \ldots, S_n(\theta))$ together with the integers $1, 2, \ldots, n$.

Alternatively, it may be noted that the basic permutation argument of Section 2.3.1 can be employed to obtain the conditional distribution of W after replacing the values $|x_i - \theta|$ by their ranks. Since the set of ranks is the same for all samples, this conditional distribution is also the unconditional distribution.

Example 2.9 With $n = 3$ have the following tabulation:

$R_i(\theta)$	$S_i(\theta)$ [only signs are indicated]							
1	+	−	+	+	−	−	+	−
2	+	+	−	+	−	+	−	−
3	+	+	+	−	+	−	−	−
W	6	4	2	0	0	−2	−4	−6

This gives the distribution

$$w : \quad -6 \quad -4 \quad -2 \quad 0 \quad 2 \quad 4 \quad 6$$
$$8\Pr[W = w] : \quad 1 \quad 1 \quad 1 \quad 2 \quad 1 \quad 1 \quad 1$$

The null distribution of W has been tabulated for certain values of n; for example, Lehmann (1975), Table H. Some care should be exercised when using tables, because some authors tabulate the distribution of W_+, the sum of the positive ranks; $W = 2W_+ - n(n + 1)/2$. The null distribution of W does not depend on F, hence inference based on W is distribution-free, except for the assumption of symmetry.

The distribution of W is clearly symmetric about 0, hence $E\{W(\mathbf{X}, \theta)\} = 0$. Conditionally on fixed $(R_1(\theta), \ldots, R_n(\theta))$, W is a weighted sum of independent random variables $S_i(\theta), i = 1, 2, \ldots, n$, each of which has variance 1, the weights being the integers $1, 2, \ldots, n$. Hence,

$$\text{var}\,\{W(\mathbf{X}, \theta) | \mathbf{R}(\theta)\} = \sum_{i=1}^{n} i^2 = n(n + 1)(2n + 1)/6 \qquad (2.24)$$

and since $E\{W(\mathbf{X}, \theta) | \mathbf{R}(\theta)\} = 0$, using $\text{var}(W) = E\{\text{var}(W|\mathbf{R}\}$ $+ \text{var}\,\{E(W|R)\}$ shows that the unconditional var $\{W(\mathbf{X}, \theta)\}$ is also as given in (2.24). For large n the conditional distribution of W is approximately normal, by an application of Theorem 1.3. Since the conditional distribution of W is identical for all realizations of $|X_1 - \theta|, |X_2 - \theta|, \ldots, |X_n - \theta|$, the unconditional distribution of W is also approximately normal for large n.

Example 2.10 For $n = 6$, $\Pr(W \leqslant 18) = 1 - 7/128 = 0.945$.

Normal approximation: $\Pr(W \leqslant 18) \simeq \phi\left(\dfrac{18 + 1 - 0}{\sqrt{140}}\right) = 0.946$.

Note the 'continuity correction' $(+ 1)$ in the numerator of the argument for ϕ.

Testing $H_0 : \theta = \theta_0$ against $H_1 : \theta > \theta_0$

As a function of t, the statistic $W(\mathbf{x}, t)$ is non-increasing in t. Clearly, if $t < x_{(1)}$ we have $W = n(n + 1)/2$. As t increases, W changes only if $\text{sgn}(x_i - t)$, or $\text{Rank}(|x_i - t|)$, or both change. These events occur, respectively, when $t = x_i$, or when $|x_i - t| = |x_j - t|$ for some $i \neq j$. In the latter case we have $x_i - t = -x_j + t$, or $t = (x_i + x_j)/2$. In both cases W decreases by 2. At $t = x_i$ this happens because $\text{sgn}(x_i - t)$ $\text{Rank}(|x_i - t|)$ changes from $+ 1$ to $- 1$. In the second case, if $x_i \leqslant x_j$ the contribution of x_i and x_j to W is

$$-r + (r + 1) \qquad \text{for } t < (x_i + x_j)/2$$
$$-(r + 1) + r \qquad \text{for } t > (x_i + x_j)/2$$

giving a change of $- 2$. We can summarize these statements in

$$W(\mathbf{x}, t) = n(n + 1)/2 - 2 \,\#\,\{(x_i + x_j)/2 \leqslant t; \quad i, j = 1, 2, \ldots, n\} \qquad (2.25)$$

the r.h.s. being clearly non-increasing in t.

In the light of the result above we have

$$E\{W(\mathbf{X}, \theta_0)\} > E\{W(\mathbf{X}, \theta)\} \qquad (2.26)$$

hence the test procedure is

'reject H_0 if observed $W(\mathbf{x}, \theta_0) > C$'

where C is obtained by reference to the null distribution of W.

Example 2.11 Suppose that we have the $n = 10$ observations given in Example 2.1, and that $\theta_0 = 4.5$. The values of $x_i - 4.5$ and the signed ranks are listed below.

$x_i - 4.5 : -1.12, \quad 1.31, \quad -0.04, \quad 0.12, \quad -0.35, \quad 0.94, \quad 2.06, \quad 1.32,$
$\qquad\quad -0.55, \quad 0.69$

signed ranks : $-7 \qquad 8 \qquad -1 \qquad 2 \qquad -3 \qquad 6 \qquad 10 \qquad 9$
$\qquad\qquad\quad -4 \qquad 5$

$$W(\mathbf{x}, 4.5) = 25$$

For the null distribution of W,

$$\mathrm{var}(W) = 385 = (19.62)^2$$

$$\Pr\{W \geqslant 25\} \simeq 1 - \Phi(24/19.62) = 0.11$$

which is the approximate significance level.

To check the consistency of the test we rewrite (2.25) as

$$W(\mathbf{x}, t) = N(n+1)/2 - \sum_{i,j=1}^{n} V_{ij}(t)$$

where $V_{ij}(t) = \begin{cases} 1 & \text{if } (x_i + x_j)/2 \leqslant t \\ 0 & \text{otherwise} \end{cases}$.

Thus

$$E\{W(\mathbf{X}, t)\} = n(n+1)/2 - \sum_{i,j=1}^{n} E\{V_{ij}(t)\}$$

where $E\{V_{ij}(t)\} = \Pr\{(X_i + X_j)/2 \leqslant t\}$

$$= \begin{cases} F(t) & \text{when } i = j \\ \displaystyle\int_{-\infty}^{\infty} F(2t - x)f(x)\,\mathrm{d}x & \text{when } i \neq j \end{cases}$$

giving.

$$E\{W(\mathbf{X}, t)\} = n(n+1)/2 - 2nF(t)$$

$$- n(n-1) \int_{-\infty}^{\infty} F(2t - x)f(x)\,\mathrm{d}x \qquad (2.27)$$

Combining (2.26) and (2.27) we see that as $n \to \infty$,

$$E\{W(\mathbf{X}, \theta_0)\} \simeq An^2, \qquad A > 0,$$

while $\mathrm{var}\{W(\mathbf{X}, \theta_0)\} \leqslant Bn^3$, where B is some constant. Thus, from Section 1.3, the test is consistent.

Confidence limits for θ

We again use the procedure of inverting the hypothesis-testing argument. From the null distribution of W we find w such that $\Pr[|W(\mathbf{X}, \theta)| \leqslant w] = 1 - \alpha$. We have seen that as t varies from $-\infty$ to $+\infty$ the value of $W(\mathbf{x}, t)$ changes from $N = n(n+1)/2$ to $-N$ in N jumps of size 2 at points $\xi_{(1)}, \xi_{(2)}, \ldots, \xi_{(N)}$, where the $\xi_{(k)}$ are the N averages $(x_i + x_j)/2$, $i, j = 1, 2, \ldots, n$, arranged in increasing order of magnitude. The $100(1 - \alpha)\%$ two-sided confidence limits are $\xi_{(r_1)}, \xi_{(r_2)}$, where $r_1 = (N - w)/2$, $r_2 = N - r_1 + 1$.

Example 2.12 Use the following $n = 6$ observations:

$$3.38, \ 5.81, \ 4.46, \ 4.62, \ 4.15, \ 5.44$$

There are 21 pairwise averages as shown below, with the original observations underlined. The numbers shown in brackets are the values of W for t lying between the corresponding averages.

(21) 3.38	(19) 3.765	(17) 3.92	(15) 4.00
(13) <u>4.15</u>	(11) 4.305	(9) 4.385	
(7) 4.41	(5) <u>4.46</u>	(3) 4.54	(1) 4.595
(−1) <u>4.62</u>	(−3) 4.795	(−5) 4.95	
(−7) 4.98	(−9) 5.03	(−11) 5.135	(−13) 5.215
(−15) <u>5.44</u>	(−15) 5.625	(−19) <u>5.81</u>	(−21)

From the null distribution of W, $\Pr[|W| \leqslant 15] = 1 - 6/64 = 0.906$. Thus by inspection of the table above, a 90.6% confidence interval for θ is: (3.92, 5.44). This result happens to coincide with the result in Example 2.8; this will not always happen.

Note that $\mathrm{var}(W) = 91 = (9.539)^2$. Using a normal approximation for the distribution of W,

$$\Pr[|W| \leqslant 15] = 2\phi(16/9.539) - 1 = 2\phi(1.677) - 1 = 0.906.$$

Thus, using this approximation for an approximately 90% confidence interval, the result given by the exact distribution would have been obtained.

Point estimation based on W

The estimating equation for θ is

$$W(\mathbf{x}, t) = \sum_{i=1}^{n} \text{sgn}(x_i - t) \, \text{Rank}(|x_i - t|) = 0 \qquad (2.28)$$

The resulting estimate, $\hat{\theta}$, is usually called the Hodges–Lehmann estimate (Hodges and Lehmann, 1963).

Recalling that as t varies from $-\infty$ to $+\infty$, $W(\mathbf{x}, t)$ decreases from $+N$ to $-N$ in N steps of size 2 at the values $\xi_{(1)}, \xi_{(2)}, \ldots, \xi_{(N)}$, we see that (2.28) may not, strictly, have a unique solution. However, we take the solution to be the median of the $\xi(k)$ values; thus the solution is defined as

$$\hat{\theta} = \begin{cases} \xi_{((N-1)/2 + 1)} & \text{if } N \text{ is odd} \\ \{\xi_{(N/2)} + \xi_{(N/2 + 1)}\}/2 & \text{if } N \text{ is even} \end{cases}$$

Owing to the symmetry of X, the estimate $\hat{\theta}$ is median-unbiased for θ.

Example 2.13 In Example 2.12 the point estimate of θ produced from (2.28) is $\hat{\theta} = 4.595$.

The point estimate $\hat{\theta}$ is consistent for θ because we have seen from (2.27) that, as $n \to \infty$,

$$E\{W(\mathbf{X}, t)/n\} \simeq A$$

While

$$\text{var}\{W(\mathbf{X}, t)/n\} \leqslant B/n$$

consistency follows from Section 1.4.

According to Section 1.6.1 the distribution of $\hat{\theta}$ tends to normality and to find its large sample variance we need $[\partial E\{W(\mathbf{X}, t)\}/\partial t]_{t=\theta}$. To calculate the derivative it is convenient and entails no loss of generality to take $\theta = 0$. We can then use (2.27), and noting that $f(-x) = f(x)$, straightforward differentiation gives

$$[\partial E\{W(\mathbf{X}, t)\}/\partial t]_{t=\theta} = -2nf(0) - 2n(n-1)\int f^2(x)\,\mathrm{d}x \qquad (2.29)$$

Writing $\int f^2(x)\,\mathrm{d}x = \bar{f}$, the 'mean density', we have, using (2.24),

$$\text{var}(\hat{\theta}) \simeq 1/(12n\bar{f}^2) \qquad (2.30)$$

The efficiency of W

From (2.29) and (2.14) the efficacy of W is readily seen to be

$$e_W(\theta) = 2\sqrt{3}\bar{f} \qquad (2.31)$$

It is interesting to note that the efficacy of W depends on the mean density, $\bar{f}$, while the efficacy of S depends, by contrast, on the density at the median, $f(0)$. For many standard distributions calculation of $\bar{f}$ is easy because $f^2(x)$ is proportional to a density of the same type as $f(x)$. For example, if $f(x)$ is a normal density, $f^2(x)$ is proportional to a normal density. If $f(x)$ is a Cauchy density, both $f(x)$ and $f^2(x)$ are, apart from constants, Pearson Type VII densities.

In Table 2.1 some values of $e_W(\theta)$ are given, along with $e_S(\theta)$ and $e_{ML}(\theta)$ for comparison. Note that the large-sample variances of the corresponding point estimates are obtainable as $[n^{1/2}e_W(\theta)]^{-2}$, etc.

Table 2.1 *Efficacies of Wilcoxon signed-rank(W), median (S), mean (A), and ML statistics.*

	Wilcoxon signed-rank (W)	Median (S)	Mean (A)	ML
$N(0, 1)$	0.98	0.80	1.00	1.00
Cauchy $(0, 1)$	0.55	0.64	0	0.71
Double exponential $f(x, \theta) = \frac{1}{2}e^{-|x - \theta|}$	0.87	1.00	0.71	1.00
$[18\phi(x - \theta) + 2\phi\{(x - \theta)/10\}]/20$	0.81	0.73	0.30	0.91

Table 2.1 shows that the efficacy of W is remarkably close to that of the ML statistic in the normal case. In every example in the table, the better of W and S has satisfactory efficacy. Further, each of W and S have reasonable robustness in being reasonably efficient over a class comprising several types of distributions.

Table 2.1 also draws attention to the problem, not unique to distribution-free procedures, of choice between statistics. According to formulae (2.15) and (2.31), W is better than S when $\sqrt{3}\bar{f}/f(0) > 1$, but, even within the class of Pearson Type VII distributions, the ratio $\sqrt{3}\bar{f}/f(0)$ varies from $\sqrt{3/4}$ to $\sqrt{3/2}$ as m varies between 1 (Cauchy) and ∞ (normal). Also, examination of realized frequency distributions, based on reasonably large samples from any of the populations represented in the table, will show that it is far from easy to choose between S and W. To do so according to the criterion $\sqrt{3}\bar{f}/f(0)$, one has to replace $\bar{f}$ and $f(0)$ by estimates, and estimating these quantities is not straightforward. A better procedure may be to base the choice of procedure on a more direct estimate of efficacy, some further discussion of this topic is given in Section 2.3.7.

A further question that arises is whether some other transformation, based on ranks, might produce a better statistic than W. Heuristically, one of the reasons for the rank transformation being successful is that it shrinks all observations onto the $(0, 1)$ interval (after division by, say, $n + 1$). In the case of long-tailed distributions the effect of very large or very small observations is thus diluted. However, in some cases this effect may be too severe, and the question is whether, by manipulating the ranks, transformations can be made that would be nearly as good as ML for certain distributions. That this is possible is suggested by considering the example of 'normal scores' defined as follows: Suppose that Rank $(|x_i - t|)$ in (2.28) is r. Then replace Rank $(|x_i - t|)$ by $E(Y_{(r)})$, where Y has the standard half-normal distribution with p.d.f. $2\phi(y)$, $y \geqslant 0$. A transformation having a very similar effect is to replace Rank $(|x_i - t|)$ by $\phi_*^{-1}\{r/(n + 1)\}$, where $\phi_*(y) = 2\{\phi(y) - \tfrac{1}{2}\}$ is the c.d.f. of Y. The explanation of the similar effects of these two transformations is that $E(Y_{(r)}) = \phi_*^{-1}$ $\{r/(n + 1)\} + O(1/n)$; see, for example, David (1970, p. 65).

Example 2.14 Consider the following $n = 9$ observations and their transforms:

x_i	1.39	-0.56	0.05	0.32	2.94	1.97	-0.26	0.41	0.44
$r_i = \text{Rank}(|x_i|)$	7	6	1	3	9	8	2	4	5
$\text{sgn}(x_i)E(Y_{(r_i)})$	1.09	-0.88	0.13	0.39	1.84	1.37	-0.26	0.54	0.70
$\text{sgn}(x_i)\phi_*^{-1}\{r_i/(n + 1)\}$	1.04	-0.84	0.12	0.38	1.64	1.28	-0.25	0.53	0.67

Note that the scores give a reasonable reproduction of the original observations; also note the closeness of the two sets of scores. The agreement between scores and original observations suggests that, in the normal case, procedures based on the scores could have efficiencies close to that of the likelihood method. The following sections pursue the question of scores somewhat further.

2.3.4 *Scores based directly on ranks*

We generalize the statistic $W(\mathbf{X}, t)$ defined in (2.23) by replacing $R_i(t) = \text{Rank}(|X_i - t|)$ by $G_i(t) = G\{R_i(t)/(n + 1)\}$, where $G(u)$ is continuous, monotonic increasing in u, and defined on $(0, 1)$.

Typically G^{-1} is a distribution function, as in Example 2.14. Observe that when $G^{-1}(u) = u$, the c.d.f. of a uniform distribution on $(0, 1)$, then $G_i(t) = R_i(t)/(n + 1)$. However, G need not always be an inverse distribution function, as we shall see in Section 2.3.5.

We now have the statistic

$$W_G(\mathbf{X}, t) = \sum_{i=1}^{n} \text{sgn}(X_i - t) G_i(t) \qquad (2.32)$$

of which $W(\mathbf{X}, t)$ is clearly a special case. Note that for a given n there are only n distinct scores $a_i = G\{i/(n + 1)\}$, $i = 1, 2, \ldots, n$, and every vector $\mathbf{G}(t) = (G_1(t), G_2(t), \ldots, G_n(t))$ is a permutation of these a_i.

The null distribution of W_G

Owing to the independence of the vectors $\mathbf{S}(\theta) = (S_1(\theta), S_2(\theta), \ldots, S_n(\theta))$ and $\mathbf{R}(\theta) = (R_1(\theta), R_2(\theta), \ldots, R_n(\theta))$, noted in Section 2.3.3, the vectors $\mathbf{S}(\theta)$ and $\mathbf{G}(\theta)$ are also independent. Hence the null distribution of W_G can be tabulated by exactly the same process of enumeration used for the null distribution of W. This distribution is symmetric about 0 hence $E\{W_G(\mathbf{X}, \theta)\} = 0$, and following the argument leading to (2.24),

$$\text{var}[W_G(\mathbf{X}, \theta)] = \sum_{i=1}^{n} G_i^2(t) \qquad (2.33)$$

The distribution of $W_G(\mathbf{X}, \theta)$ can be approximated by a normal distribution for large n if $\max[G_i^2(t)]/\sum G_i^2(t) \to 0$ as $n \to \infty$. For large n it is sometimes useful to approximate the sum in (2.33) by an integral giving

$$\text{var}[W_G(\mathbf{X}, \theta)] \simeq n \int_0^1 G^2(u) \, du \qquad (2.34)$$

Note that in the case $G^{-1}(u) = u$, giving $W_G(\mathbf{X}, \theta) = W(\mathbf{X}, \theta)/(n + 1)$, equation (2.34) gives $\text{var}[W(\mathbf{X}, \theta)] \simeq (n + 1)^2 n/3$; asymptotically this result and the result in (2.24) are identical.

Testing $H_0 : \theta = \theta_0$

The testing procedure is straightforward and is illustrated by Example 2.15 below. Since $G(u)$ is monotonic in u, consistency of a test based on W_G for a one-sided alternative $H_1 : \theta > \theta_0$ follows by arguments similar to those of Section 2.3.3 for W. Modifications for other H_1 are obvious.

Example 2.15 Suppose we have the $n = 6$ observations given below from a symmetric population and wish to test $H_0 : \theta = 0$ against $H_1 : \theta > 0$ at level $\alpha \simeq 0.10$.

x_i	-0.89	5.28	-1.32	2.21	0.56	0.50
Rank $|x_i|$	3	6	4	5	2	1
Scores	-0.56	1.46	-0.83	1.07	-0.37	-0.18

The scores are obtained from $G^{-1}(u) = 2 \int_0^u \phi(x)\,dx$ as in Example 2.14. The six largest values of W_G occurring with equal probabilities are obtained with the following signs applied to the transformed ranks arranged in increasing order of magnitude.

	0.18	0.37	0.56	0.83	1.07	1.46	W
	+	+	+	+	+	+	4.47
	−	+	+	+	+	+	4.11
	+	−	+	+	+	+	3.73
	−	−	+	+	+	+	3.37
	+	+	−	+	+	+	3.35
	−	+	−	+	+	+	2.99
observed	−	−	−	−	+	+	0.60

$\Pr[W_G \geqslant 2.99] = 6/64$, observed $W_G = 0.60$, thus H_0 is accepted at level $\alpha = 6/64 = 0.94$.

Confidence limits for θ, and point estimation
By reference to the null distribution of W_G, w is found such that $\Pr[|W_G(\mathbf{X}, \theta)| \leqslant w] = 1 - \alpha$, with $1 - \alpha$ close to the selected confidence coefficient. Then the values of $W_G(\mathbf{x}, t)$ are calculated for t varying between $-\infty$ and $+\infty$, noting that $W_G(\mathbf{x}, t)$ varies between $\sum G_i(t)$ and $-\sum G_i(t)$ with $n(n+1)/2$ jumps, of unequal size, at the averages $(x_i + x_j)/2$, $i, j = 1, 2, \ldots, n$. The confidence limits for θ can then be determined by inspection of this tabulation to find the θ values satisfying $|W_G(\mathbf{x}, \theta)| \leqslant w$.

Example 2.16 Use the data and scoring function in Example 2.15. Set $\alpha = 6/64$ for a two sided 90.6% confidence interval. Using the tabulation in Example 2.15 we find that $\Pr[|W_G(\mathbf{X}, \theta)| \leqslant 3.37] = 1 - 6/64$. The $n(n+1)/2 = 21$ averages $(x_i + x_j)/2$ are listed in order below, with the values of $W_G(\mathbf{x}, t)$ for t in the relevant intervals shown in brackets; the original observations are underlined.

(4.47)	-1.32	(4.11)	-1.105	(3.73)	-0.89	(3.37)	-0.41	(2.99)
(2.99)	-0.38	(2.45)	-0.195	(2.07)	-0.165	(1.69)	0.445	(1.21)
(1.21)	0.50	(0.85)	0.53	(0.47)	0.56	(0.11)	0.66	(-0.43)
(-0.43)	1.355	(-0.81)	1.385	(-1.19)	1.98	(-1.97)	2.195	(-2.45)
(-2.45)	2.21	(-2.81)	2.89	(-3.35)	2.92	(-3.73)	3.745	(-4.11)
(-4.11)	5.28	(-4.47)						

From the table, the interval of t values for which $|W_G(\mathbf{x}, t)| \leqslant 3.37$ is $(-0.89, 2.92)$.

Formally the point estimate of θ is the value of t at which $W_G(\mathbf{x}, t) = 0$. In general $W_G(\mathbf{x}, t)$ jumps from a positive value, a, to a negative value, $-b$, at one of the $(x_i + x_j)/2$ averages, $\hat{t}$ say, with $a \neq b$. If $a = b$ it would be natural to take the solution $\hat{\theta}_G = \hat{t}$ but since this does not happen in general it might be more satisfactory to 'smooth' the graph of $W_G(\mathbf{x}, t)$ against t in the neighbourhood of $W_G(\mathbf{x}, t) = 0$ and thus make an interpolation to find t_G. For the data of Example 2.16 a segment of the graph of $W_G(\mathbf{x}, t)$ against t is shown in Fig. 2.1. The smoothing shown in this figure suggests taking $\hat{\theta}_G = 0.63$ rather than $\hat{t} = 0.66$.

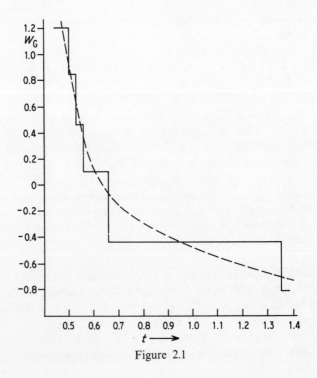

Figure 2.1

Consistency of $\hat{\theta}_G$ can be established by arguments similar to those used for $\hat{\theta}$. The large sample variance of $\hat{\theta}_G$ can be obtained from the efficacy of W_G, a discussion of which follows.

Efficiency of W_G

In Section 2.3.3 it was possible to derive an exact expression for $E\{W(\mathbf{X}, t)\}$ by simple steps because of the possibility of expressing Rank$(|Z_i(t)|)$ as a sum of indicator random variables. This simple procedure is unavailable when we consider $G[\text{Rank}(|Z_i(t)|)/(n+1)]$. However, we can make use of the device of replacing Rank$(|Z_i(t)|/(n+1))$ by $F^*(|Z_i(t)|)$, where $F^*(\ )$ is the c.d.f. of $|Z_i(t)|$, invoking the following lemma, whose proof is obvious.

Lemma 2.1 Suppose that $G(u)$ is continuous and twice differentiable in $(0, 1)$ with $0 < G(u) < M$. If $X_1, X_2, \ldots, X_n$ are identically and independently distributed with c.d.f. $F(x)$, then

$$E\{\text{Rank}(X_i)/(n+1)|X_i = x\} = (n-1)F(x)/(n+1) + O(1/n) \simeq F(x)$$

$$\text{var}\{\text{Rank}(X_i)/(n+1)|X_i = x\} = (n-1)F(x)\{1 - F(x)\}/(n+1)^2$$

Also

$$E[G\{\text{Rank}(X_i)/(n+1)\}|X_i = x] \simeq G\{F(x)\} + O(1/n)$$

Let us now consider an alternative derivation to that given in Section 2.3.3 of $E\{\sum_{i=1}^n \text{sgn}(X_i - t)\,\text{Rank}(|Z_i(t)|)\}$. It is convenient without loss of generality, to take $\theta = 0$. Then the p.d.f. of $(X - t)$ is $f(x + t)$ and the p.d.f. of $Y = |X - t|$ is

$$f^*(y, t) = f(y + t) + f(-y + t), \qquad \text{for } y \geqslant 0$$

with c.d.f.

$$F^*(y, t) = \int_0^y [f(u + t) + f(-u + t)]\,du \qquad (2.35)$$

Also

$$\Pr\{\text{sgn}(X - t) = +1 | |X - t| = y\}$$

$$= f(y + t)/\{f(y + t) + f(-y + t)\} \qquad (2.36)$$

Conditioning on $\mathbf{Y} = (Y_1, Y_2, \ldots, Y_n) = (y_1, y_2, \ldots, y_n) = \mathbf{y}$ we find, since the conditioning fixes the ranks of the $|Z_i(t)|$, and taking expectations with respect to the $\text{sgn}(X_i - t)$, random variables

$$E\{W(\mathbf{X}, t) | \mathbf{Y} = \mathbf{y}\} = \sum_{i=1}^{n} \frac{f(y_i + t) - f(-y_i + t)}{f(y_i + t) + f(y_i + t)} \text{Rank}(y_i)$$

$$= \sum_{i=1}^{n} Q(y_i, t) \tag{2.37}$$

Next consider only the term $Q(y_j, t)$ in the sum in (2.37) and condition only on $Y_j = y$, while allowing the other y_i values to vary. Then, using Lemma 2.1, and neglecting the term of order $1/n$,

$$E\{Q(Y_j, t) | Y_j = y\} = (n-1) \frac{f(y + t) - f(-y + t)}{f(y + t) + f(-y + t)} F^*(y, t) \tag{2.38}$$

The unconditional expectation of this term is obtained by integrating out with respect to the p.d.f. of Y and we obtain

$$E\{Q(Y_j, t)\} = (n-1) \int_{0}^{\infty} [f(y + t)$$

$$-f(-y + t)]F^*(y, t)\,dy \tag{2.39}$$

Using (2.35) and noting that $f(-a) = f(+a)$, and $f'(-a) = -f'(a)$, we find after differentiating the expression in (2.39) with respect to t and putting $t = 0$, that

$$[\partial E\{W(\mathbf{X}, t)\}/\partial t]_{t=0} \simeq 2n^2 \int_{-\infty}^{\infty} f^2(y)\,dy \tag{2.40}$$

in agreement with (2.29), as $n \to \infty$.

Replacing $\text{Rank}\{|Z_i(t)|\}$ by $G\{\text{Rank}(|Z_i(t)|)/(n+1)\}$, we use the same conditioning arguments, and the latter part of Lemma 2.1 in the step (2.37) to (2.38), to obtain

$$E[W_G(\mathbf{X}, t)] \simeq n \int_{0}^{\infty} [f(y + t) - f(-y + t)]G(F^*(y, t))\,dy$$

and finally

$$[\partial E\{W_G(\mathbf{X}, t)\}/\partial t]_{t=0} \simeq - \int_{0}^{\infty} 4f^2(u)G'[2F(u) - 1]\,du \tag{2.41}$$

Combining (2.34) and (2.41) we get

$$e_{W_G}(\theta) = \left\{ 4 \int_0^\infty f^2(u) G'[2F(u) - 1] \, du \right\} \Bigg/$$

$$\left\{ \int_0^1 G^2(u) \, du \right\}^2 \tag{2.42}$$

Formula (2.42) is, of course, useful for finding the efficacy of a test based on any transformation G, and several special cases have been studied. For example, if F is a normal distribution and G is the 'normal scores transformation', it can be verified that the ARE of W_G and the MLE is 1.

2.3.5 Optimum rank statistics

In practice it is not at all easy to decide on a 'best' G, a matter that has been discussed before in Section 2.3.4. If there are grounds for thinking that the underlying distribution might be normal, then obviously a normal-scores transformation would be selected. However, perhaps the most useful aspect of the efficiency studies in this section is that one is enabled to study the efficiency of a particular type of score over a variety of distributions F; Chernoff and Savage (1958) have shown, for example, that the normal-scores test has a minimum asymptotic efficiency of 1 relative to the usual t-test over all distributions F, suggesting the former to be always preferable to the latter.

In the discussion of normal scores in Section 2.3.3, it was pointed out that when sampling from a normal population the scores tend to be quite close to the original observations, because $Y_{(r)}$ is an unbiased estimate of $E(Y_{(r)})$ with variance of order $1/n$. Further, in the normal case the ML estimating equation happens to be simply

$$\sum_{i=1}^n \text{sgn}(X_i - t)|X_i - t| = 0 \tag{2.43}$$

Thus a natural question for a general symmetric distribution is whether a score can be defined such that a statistic based on such scores behaves like the likelihood statistic.

For symmetric $F(x, \theta) = F(x, -\theta)$, the ML estimating equation in the 'regular' case can be written

$$\sum_{i=1}^{n} \text{sgn}(x_i - t)f'(|x_i - t|)/f(|x_i - t|) = 0$$

Hence, using the normal case as a guide, we define scores

$$a(i) = E\{f'(Y_{(i)})/f(Y_{(i)})\} \qquad (2.44)$$

It will be noted that since the $Y_{(i)}$ are the order statistics of the random variables $|X_1 - \theta|, |X_2 - \theta|, \ldots, |X_n - \theta|$, the values of the $a(i)$ can be tabulated once and for all for a given F and n. Using these scores, the estimating equation is

$$\sum_{i=1}^{n} \text{sgn}(x_i - t)a(R_i) = 0 \qquad (2.45)$$

where R_i is the rank of $|x_i - t|$.

Since $E(Y_{(r)}) = F_*^{-1}\{r/(n+1)\} + O(1/n)$, $r = 1, 2, \ldots, n$, where $F_*(u) = 2\{F(u) - \frac{1}{2}\}$ (see, for example, David, 1970, p. 65), the score in (2.44) can be approximated by

$$a^*(i) = f'[F_*^{-1}\{i/(n+1)\}]/f[F^{-1}\{i/(n+1)\}] \qquad (2.46)$$

Using the scores $a^*(i)$, and formula (2.42), the efficacy of the resulting statistic can be shown to equal the efficacy of the ML statistics (Terry, 1952).

In potential applications of rank methods, the usefulness of scoring systems such as (2.44) and (2.46) is problematical, because rank methods are normally considered in precisely those situations where there is uncertainty about the form of F. However, they can be useful for diluting the effect of outlying observations. Asymptotically little efficiency may be lost by using a statistic based on one of the scores, and there may be some gain in *robustness*.

2.3.6 *Robust transformations*

As we have remarked in Section 2.3.4, the success of some of the rank methods in coping with certain long-tailed distributions is attributable to their shrinking effect on outlying observations. Other transformations of the observations to achieve a similar effect are possible. Two types of transformations that we shall discuss rather briefly are related to *M*-estimates and *L*-estimates.

M-estimates
Suppose that we transform $x_i - t$ to $\psi(x_i - t)$, where $\psi(u)$ is con-

tinuous in u and differentiable almost everywhere. Then consider the 'M-statistic'

$$M_\psi(\mathbf{X}, t) = \sum_{i=1}^{n} \psi(X_i - t) \tag{2.47}$$

see also (2.6). When X is symmetrically distributed it is natural to choose ψ such that $\psi(-u) = -\psi(u)$; to ensure this we use instead

$$M_\psi(\mathbf{X}, t) = \sum_{i=1}^{n} \operatorname{sgn}(X_i - t)\psi(|X_i - t|) \tag{2.48}$$

In this form we obviously have

$$E\{M_\psi(\mathbf{X}, t)\} = 0$$

leading to the type of estimating equation for θ that we have discussed before.

As an example of $\psi(u)$, take $\psi(u) = u/(1 + u^2)$. Then it will be noticed that the estimating equation for θ is exactly the equation that is obtained by the ML method if the p.d.f. of X is the Cauchy $f(x, \theta) = \pi^{-1}[1 + (x - \theta)^2]^{-1}$.

The fact that estimates based on M_ψ may coincide with ML estimates has inspired the terminology 'M-estimates'.

L-estimates
These are linear functions of the order statistics $X_{(1)} < X_{(2)} < \ldots < X_{(n)}$ in the form

$$t_L = \sum_{i=1}^{n} W_i X_{(i)}$$

where $\sum_{i=1}^{n} W_i = 1$; t_L is the solution of the estimating equation

$$L(\mathbf{X}, t) = \sum_{i=1}^{n} W_i(X_{(i)} - t) = 0. \tag{2.49}$$

If $n = 2k + 1$, $w_k = 1$, $w_j = 0$ for $j \neq k$ we obtain the estimating equation (2.8) as a special case of (2.49).

Tests and confidence limits based on M-statistics.
To test $H_0 : \theta = \theta_0$ the basic permutation argument can be applied in an obvious way to $M_\psi(\mathbf{X}, t)$, written as in (2.38). Further, the same argument can be used to find *exact* confidence intervals for θ.

Determination of confidence limits is more tedious than with the rank methods because the null conditional distribution of $M_\psi(\mathbf{X}, \theta)$ is not invariant with respect to θ. In practice a null distribution must be found for every trial θ and the observed M_ψ referred to it; this applies also, of course, to the mean statistic A.

Following the argument used in connection with the mean statistic A, a $100(1 - 2r/2^n)\%$ two-sided confidence interval for θ is (t_1, t_2), where

$t_1(t_2)$ is the rth smallest (largest) t such that

$$\sum_{i=1}^{s} \psi(|x_i' - t|) = 0$$

where for $1 \leqslant s \leqslant n$, $x_1', x_2', \ldots, x_s'$ is a subset of the sample $x_1, x_2, \ldots, x_n$.

Example 2.17 Use the $n = 6$ observations of Example 2.15; they are shown in the tabulation below. Take $\psi(u) = (1 - e^{-u})/(1 + e^{-u})$, following a suggestion of Huber (1972). We shall find a $100(1 - 6/64)\%$ two-sided confidence interval for θ.

To decide whether a trial value t of θ belongs to the confidence interval, we first find the $n = 6$ values of $\psi(|x_i - t|)$ and enumerate the conditional distribution of M_ψ by listing the 2^n sign combinations. In fact, we only need the three largest (and smallest) possible values of M_ψ for our present purpose; note that the distribution of M_ψ is symmetrical about 0. If observed M_ψ lies between its third largest and third smallest possible value, t belongs to the confidence interval. For example, take $t = -0.8$ and $t = 2.92$ to give the following results:

x_i	sgn $(x_i - t)\psi$ $(|x_i - t|)$ $t = -0.8$	$t = 2.92$
-1.32	-0.254	-0.972
-0.89	-0.045	-0.957
0.50	0.572	-0.837
0.56	0.592	-0.827
2.21	0.906	-0.341
5.28	0.995	0.827
Observed M_ψ:	2.766	-3.933
3rd largest M_ψ:	2.856	3.933

The table above shows that the value $t = -0.8$ is in the confidence interval; likewise $t = 2.92-$; $t = 2.92+$ is outside the interval.

Straightforward calculations show that the confidence interval is $(0.89, 2.92)$. This result coincides with the result of Example 2.16; generally this will not happen.

When n is large, judicious choice of $\psi(u)$ will ensure that the conditional null distribution of M_ψ can be approximated by a normal distribution. Thus, with $\psi(u) = (1 - e^{-u})/(1 + e^{-u})$, $u \geqslant 0$, $\psi(u)$ is bounded between 0 and 1 so that a normal approximation is possible. When a normal approximation can be used, the confidence limits are given by the solutions of

$$\sum_{i=1}^{n} \text{sgn}(x_i - t)\psi(|x_i - t|) = \pm u_\alpha \left\{ \sum_{i=1}^{n} [\psi(|x_i - t|)]^2 \right\}^{1/2} \tag{2.50}$$

where u_α denotes an appropriate normal deviate; the solutions are generally fairly easy to obtain numerically.

Example 2.18 Applying the normal approximation in (2.50) to the data in Example 2.17, with $u_\alpha = 1.645$ for an approximately 90% confidence interval gives the result $(-0.70, 3.22)$. This agrees reasonably well with the exact result considering that n is rather small.

One of the much studied types of robust transformations is

$$\psi(|u|) = \begin{cases} |u| & \text{for } |u| < k \\ k & \text{for } |u| > k \end{cases} \tag{2.51}$$

In principle the determination of confidence limits using (2.51) presents no new problems. However, with k relatively small, many of the $\psi(|x_i - t|)$ values may be identical, the effect of which is to reduce the number of steps in the conditional c.d.f. of M_ψ. This can be a nuisance in achieving a desired confidence coefficient. In the use of transformation (2.51), as with most robust transformations, the question of scaling arises; it is raised again in a later section.

Apart from problems already mentioned there are no serious difficulties associated with the use of transformations $\psi(u)$ that are monotonic in $u > 0$. However, the use of transformations like $\psi(u) = u/(1 + u^2)$ can lead to problems of non-existence of confidence intervals with specified confidence coefficient, especially in small samples.

M-estimates and their efficiencies
Point estimates based on M_ψ statistics are found by solving the

appropriate estimating equations. In practice it is perhaps simplest to obtain solutions graphically as indicated in the example using rank-based scores.

To find $E\{M_\psi(\mathbf{X}, t)\}$ when $\theta = 0$, the conditioning employed for equation (2.37) is useful. Conditional on $|X_i - t| = u$ we have

$$E[\operatorname{sgn}(X_i - t)\psi(|X_i - t|)||X_i - t| = u] = \frac{f(u + t) - f(-u + t)}{f(u + t) + f(-u + t)}\psi(u)$$

whence

$$E\{\operatorname{sgn}(X_i - t)\psi(X_i - t)\} = \int_0^\infty [f(u + t) - f(-u + t)]\psi(u)\,du$$

and

$$[\partial E\{M_\psi(\mathbf{X}, t)\}/\partial t]_{t=0} = 2n \int_0^\infty f'(u)\psi(u)\,du \qquad (2.52)$$

Further, since $\operatorname{sgn}(X_i - t)$, $\psi(X_i - t)$, $i = 1, 2, \ldots, n$, are independent,

$$\operatorname{var}[M_\psi(\mathbf{X}, 0)] = \int_{-\infty}^{+\infty} \psi^2(u)f(u)\,du \qquad (2.53)$$

from which, using (2.52), the efficacy of M_ψ is readily found.

M-estimation and scaling

One of the problems associated with the use of M-transformations is scaling. It will be clear that, in general, the answers produced by M-transformations are not scale-invariant. In order to achieve scale invariance the M-statistic could be written in the form

$$M_\psi(\mathbf{X}, t) = \sum_{i=1}^n \operatorname{sgn}(X_i - t)\psi[|(X_i - t)/\sigma|] \qquad (2.54)$$

where σ is a scale parameter. In practice σ will be unknown and will have to be replaced by an estimated value. (See, for example, Huber, 1972, where the alterations in efficacy calculations are mentioned.)

We shall be content to explain the use of scaling in practice only and to indicate that it is still possible to obtain exact confidence intervals, and perform exact tests. The argument here again uses conditioning on the set of observed $|x_i - t|$, $i = 1, 2, \ldots, n$, and we use an estimated scale factor $\hat{\sigma}$ that depends on the set of magnitudes $|x_i - t|$, $i = 1, 2, \ldots, n$. Writing $y_i = |x_i - t|$, $i = 1, 2, \ldots, n$, possible $\hat{\sigma}$

are

$$\hat{\sigma}_1 = [(1/n) \sum_{i=1}^{n} y_i^2]^{1/2}$$

$$\hat{\sigma}_2 = \text{median} (y_1, y_2, \ldots, y_n) \qquad (2.55)$$

The first of these two scale estimates, $\hat{\sigma}_1$, is clearly sensitive to outliers, whereas $\hat{\sigma}_2$ is more robust against outlying observations.

Hypothesis testing using M_ψ

To test $H_0 : \theta = \theta_0$ the first step, in principle, is enumeration of the null distribution of M_ψ. The values $|x_i - \theta_0|$, $i = 1, 2, \ldots n$, have to be listed and they are used to find $\hat{\sigma}$. Then the values $\psi(|x_i - \theta_0|/\hat{\sigma})$ can be listed, and the 2^n sign configurations can be applied to obtain the 2^n equi-probable values of M_ψ. In practice all possible values need not be listed; for a one-sided test at level $r/2^n$, only the r largest (or smallest) possible M_ψ values are required, as well as the observed value.

With large n, approximation of the distribution of M_ψ by a normal distribution may be possible, in which case the test of H_0 may be performed by calculating

$$\sum_{i=1}^{n} \text{sgn} (x_i - \theta_0)\psi(|x_i - \theta_0|/\hat{\sigma}) \bigg/ \left[\sum_{i=1}^{n} \{\psi(|x_i - \theta_0|/\hat{\sigma})\}^2 \right]^{1/2}$$

$$(2.56)$$

and entering a table of the standard normal distribution function. Alternatively, the desired quantile of the distribution can be approximated by $u_\alpha[\sum_{i=1}^{n} \{\psi(|x_i - \theta_0|/\hat{\sigma})\}^2]^{1/2}$, where u_α is a suitable normal deviate.

Example 2.19 $n = 10$, $H_0 : \theta = -10.0$, $H_1 : \theta \neq -10.0$, $\psi(u) = (1 - e^{-u})/(1 + e^{-u})$, $y_i = |x_i + 10|$, $\hat{\sigma}_2 = \text{median} (y_1, y_2, \ldots, y_n) = 14.55$

x_i	$|x_i + 10|$	$\psi(y_i/\hat{\sigma})$	x_i	$|x_i + 10|$	$\psi(y_i/\hat{\sigma})$
-91.6	81.6	0.993	1.1	11.1	0.364
-20.1	10.1	0.334	8.0	18.0	0.550
-7.4	2.6	0.089	9.0	19.0	0.574
-2.1	7.9	0.265	11.8	21.8	0.635
-1.3	8.7	0.290	224.0	234.0	1.000

Observed $M_\psi = 2.440$. For a two-sided test at the 10% level we need the 51st largest (smallest) possible M_ψ value $= 3.108 (-3.108)$.

For the normal approximation

$$(0.970^2 + 0.163^2 + \ldots + 1.000^2)^{1/2} = 1.851$$

giving

$$\pm 1.645 \times 1.851 = \pm 3.045$$

in reasonable agreement with the exact quantiles. Since observed $M_\psi = 2.753$ is smaller than 3.108, H_0 is accepted at the 10% level.

Point estimation and confidence limits
In principle, we can proceed exactly as in the case where no scale estimate is used. A potential difficulty arises in that $M_\psi(\mathbf{x}, t)$, with σ replaced by $\hat{\sigma}$, is no longer necessarily monotonic in t for fixed $\mathbf{x}$. With moderately large n the problem is, apparently, not serious because local maxima or minima only occur at large values of $|t|$. With some ψ the problem of monotonicity of M_ψ can be overcome by a simple modification of ψ. For example, with $\psi(u) = (1 - e^{-u})/(1 + e^{-u})$, $\hat{\sigma} = \text{median}\,(y_1, y_2, \ldots, y_n)$, we can put

$$\psi_*(u) = \begin{cases} \psi(u) & u \leqslant 1 \\ \psi(1) & u \geqslant 1 \end{cases} \tag{2.57}$$

Example 2.20 We use the $n = 10$ observations in Example 2.19. The two scales estimates $\hat{\sigma}_1$ and $\hat{\sigma}_2$ defined above were used with $\psi(u) = (1 - e^{-u})/(1 + e^{-u})$ and $\psi_*(u) = \min\{\psi(u), \psi(1)\}$, as in (2.57). The following point estimates and two-sided 90% confidence limits were obtained:

Method	Point estimate	Exact confidence limits	Normal approximation
Mean (A)	13.6	$-25.4, 58.9$	
Median (S)	3.4	$-7.4,\ 9.0$	$-7.4,\ 9.0$
Wilcoxon (W)	0.8	$-11.1, 10.4$	$-11.1, 10.4$
M_ψ:			
$\psi, \hat{\sigma}_1$	6.1	$-24.7, 48.5$	$-26.5, 48.9$
$\psi, \hat{\sigma}_2$	0.4	$-18.1, 14.2$	$-17.4, 14.3$
$\psi_*, \hat{\sigma}_1$	-0.1	$-24.7, 26.8$	$-24.2, 26.5$
$\psi_*, \hat{\sigma}_2$	1.3	$-9.5,\ 8.5$	$-9.5,\ 8.2$

2.3.7 *Large-sample calculations*

When the sample size is large it may be impractical to perform the exact calculations outlined in preceding sections, especially in those cases where ranking of observations is needed. One may have to resort to approximate calculations on grouped data. With large sample sizes it may well be that data are only supplied in grouped form. The sample of $n = 50$ observations in Table 2.2 will be used to illustrate some calculations using grouping; in each case a point estimate of θ with 90% confidence limits will be found.

Table 2.2

3.081,	3.459,	3.527,	3.708,	3.793,	3.854,	3.865,	3.876,	4.072,	4.097
4.204,	4.277,	4.390,	4.411,	4.434,	4.449,	4.603,	4.642,	4.685,	4.734
4.772,	4.883,	4.889,	4.975,	4.981,	5.005,	5.053,	5.138,	5.304,	5.324
5.465,	5.507,	5.544,	5.580,	5.813,	5.880,	5.995,	6.004,	6.278,	6.373
6.435,	6.449,	6.604,	6.695,	6.699,	6.701,	6.975,	7.016,	7.232,	7.666

Example 2.21 *Sign statistic*: Using the normal approximation to the null distribution of $S(\mathbf{X}, \theta)$, we have $\Pr[S(\mathbf{X}, \theta)| \leqslant 10] \simeq 2\phi(1.556) - 1 = 0.880$. Therefore the (approximately) 90% confidence limits for θ are

$$x_{(20)} = 4.734 \quad \text{and} \quad x_{(31)} = 5.465$$

and the point estimate $\hat{\theta} = (x_{(25)} + x_{(26)})/2 = 4.993$.

Table 2.3 shows the data of Table 2.2 summarized in a frequency distribution. From this distribution the values of $S(\mathbf{X}, t)$ can be found for t-values coinciding with the class end-points. The largest value of S is 50 and the smallest is -50. As t varies between $-\infty$ and $+\infty$, the reduction in S as t varies, between two class end-points is $2f_j$ where f_j is the class frequency. The values of t at which $S(\mathbf{X}, t) \simeq \pm 10$ can be found from a graph of S against t in which the tabulated S-ordinates at t-abscissa are joined by straight-line segments; note that such a graph represents the sample c.d.f. of X, except for scale and origin transformation on the ordinate. Equivalently, t for $S \simeq \pm 10$ can be found by linear interpolation. The results are 4.70, 5.42; the point estimate from the frequency distribution is $t = 5.0$. These results agree well with the ungrouped results.

Example 2.22 *Wilcoxon signed-rank statistic*: Using a normal

Table 2.3

Class	Frequency	t	S
		3.0	50
3.0–3.5	2		
		3.5	46
3.5–4.0	6		
		4.0	34
4.0–4.5	8		
		4.5	18
4.5–5.0	9		
		5.0	0
5.0–5.5	6		
		5.5	−12
5.5–6.0	6		
		6.0	−24
6.0–6.5	5		
		6.5	−34
6.5–7.0	5		
		7.0	−44
7.0–7.5	2		
		7.5	−48
7.5–8.0	1		
		8.0	−50

approximation,　$\Pr\left[\,|\,W(\mathbf{X},\theta)|\leqslant 339\right]\simeq 2\phi(340/207.183)-1=$ $2\phi(1.641)-1=0.899$. Referring to Section 2.3.4, the (approximately) 90% confidence limits for θ are

$$\xi_{(468)}=4.870 \qquad \text{and} \qquad \xi_{(808)}=5.434$$

The 1275 mean values $\xi_{ij}=(x_i+x_j)/2$, $i,j=1,2,\dots,50$, arranged in increasing order of magnitude and labelled $\xi_{(1)},\xi_{(2)},\dots,\xi_{(1275)}$, are not listed. The point estimate of θ is $\xi_{(638)}=5.142$.

An approximate procedure can be used starting with the frequency distribution in Table 2.3. The approximation is based on taking all observations in a certain class interval equal to its mid-point. Then we have possible averages 3.75, 4.00, 4.25, …, 7.75. Table 2.4 shows a convenient way of setting out the calculation of the frequency distribution of ξ_{ij} values. Each cell shows the number of occurrences of the average of the corresponding marginal values. For example, the frequency 9 in the class 4.5–5.0 of Table 2.3 generates $(9\times 10)/2=45$ averages with the value 4.75; the frequencies 8 in 4.0–4.5 and 6 in 5.0–5.5 generate $8\times 6=48$ averages at 4.75.

Table 2.5 shows a frequency distribution of ξ_{ij} values with corresponding values of W evaluated at t values mid-way between the successive mean values shown in the table. Graphically, or by linear interpolation, as in Example 2.21, the values of t at which $W\simeq\pm 339$ are

4.85 　　　 and 　　　 5.44

Table 2.4

					Class mid-points					
		3.25	3.75	4.25	4.75	5.25	5.75	6.25	6.75	7.25	7.75
	3.25	3	12	16	18	12	12	10	10	4	2
	3.75		21	48	54	36	36	30	30	12	6
	4.25			36	72	48	48	40	40	16	8
Class	4.75				45	54	54	45	45	18	9
mid-	5.25					21	36	30	30	12	6
points	5.75						21	30	30	12	6
	6.25							15	25	10	5
	6.75								15	10	5
	7.25									3	2
	7.75										1

Table 2.5

ξ_{ij}	Frequency	t	W	ξ_{ij}	Frequency	t	W
			1275	5.75	118		
3.25	3					5.875	− 771
		3.375	1269	6	86		
3.5	12					6.125	− 943
		3.625	1245	6.25	66		
3.75	37					6.375	− 1075
		3.875	1171	6.5	43		
4.0	66					6.625	− 1161
		4.125	1039	6.75	31		
4.25	102					6.875	− 1223
		4.375	835	7.0	15		
4.5	120					7.125	− 1253
		4.625	595	7.25	8		
4.75	139					7.375	− 1269
		4.875	317	7.5	2		
5	142					7.625	− 1273
		5.125	33	7.75	1		
5.25	149					7.875	− 1275
		5.375	− 265				
5.5	135						
		5.625	− 535				

and the point estimate from the graph is $t_W = 5.16$. Agreement with the ungrouped values is quite good.

Estimation of variances

Recall that the large-sample approximate variance of an estimate derived from an equation of the type $Q(\mathbf{x}, t) = 0$ involves $\{\partial E\{Q(\mathbf{x}, t)\}/\partial t\}_{t=\theta}$. The derivative $\partial E\{Q(\mathbf{x}, t)\}/\partial t$ can be estimated from the actual data by $\partial Q(\mathbf{x}, t)/\partial t$. In Examples 2.21 and 2.22, plots of $S(\mathbf{x}, t)$ and $W(\mathbf{x}, t)$ against t can be used. The slopes of these graphs are estimates of the derivatives. Estimates derived from such crude plots can possibly be refined by some mathematical smoothing process. In this connection, note that since $S(\mathbf{x}, t)$ against t is essentially the sample c.d.f. of X, the slope estimate is an estimate of the density of X; there is an extensive literature on more refined density estimation; see, for example, Rosenblatt (1971).

Example 2.23 Refer to Example 2.22; the point estimate of θ is $t_W = 5.14$ and a graph of $W(x, t)$ against t for t between 4.5 and 5.5 is shown in Fig. 2.2. From the straight line fitted by eye, shown as a broken line in the figure, the slope estimate is -1130 giving the estimated s.d. $(t_W) = \hat{\text{s.d.}}(t_W) = 0.183$. Crude 90% confidence limits for θ can be calculated using this $\hat{\text{s.d.}}(t_W)$, giving $t_W \pm 1.645\,\hat{\text{s.d.}}(t_W)$: 5.16 $\pm 0.030 = 4.86, 5.46$. These limits agree well with the limits found in Example 2.22.

2.3.8 Ties

Nothing has so far been said about ties because, theoretically, they should not occur. However, in practical measurements ties do happen, and the question arises of what to do about them when rank methods are contemplated. One simple answer is to break ties arbitrarily. For example, if measurements 2.32, 2.59, 3.41, 6.72 are made and one wishes to test $H_0 : \theta = 3$, then for $|x_i - 3|$ we have

$$0.68, \quad 0.41, \quad 0.41, \quad 3.72$$

with ranks 2 and 3 tied. The tie could be broken by considering $H_0 : \theta = 3 + \varepsilon$ or $H_0 : \theta = 3 - \varepsilon$ where $\varepsilon > 0$ is arbitrary small in magnitude.

In hypothesis testing most workers seem to favour the apparently less arbitrary device of assigning the average rank 2.5 to the two tied numbers. As far as applying the basic permutation argument is concerned, this introduces no new problem. There is only the slight

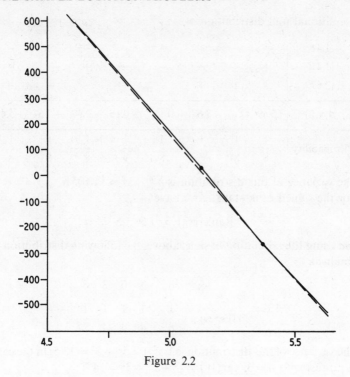

Figure 2.2

disadvantage that the simple formula for the null $\text{var}(W(\mathbf{X}, \theta))$ no longer holds.

However, when point estimation and confidence intervals are considered, where the graph of $W(\mathbf{x}, t)$ against t is used, it will be noted that this graph depicts a step function with the jumps occurring at exactly the t values which create ties, but this is no impediment to the methods that have been developed.

If two observations are actually identical, as in 2.32, 3.41, 3.41, 6.72, giving a 'real' tie rather than the 'artificial' one above, there seems no natural alternative to assigning the same rank value to the identical numbers; one method is to give the average rank of tied values to each of the tied values. Some care needs to be exercised in enumerating exact null distributions in such cases; the formula for the variance of the null distribution remains unchanged.

Example 2.24 x_i : $-2.67, 1.41, 1.41$
Test $H_0 : \theta = 0$.

Conditional null distribution of A:

1.41	+	−	+	+	−	−	+	−
1.41	+	+	−	+	−	+	−	−
2.67	+	+	+	−	+	−	−	−

$A(\mathbf{x}, 0)$	5.49	2.67	2.67	0.15	− 0.15	− 2.67	− 2.67	− 5.49
Probability	$\frac{1}{8}$	$\frac{1}{8}$	$\frac{1}{8}$	$\frac{1}{8}$	$\frac{1}{8}$	$\frac{1}{8}$	$\frac{1}{8}$	$\frac{1}{8}$

The variance of this distribution is $\sum_{i=1}^{n} x_i^2 = 11.1051$.
For the signed rank statistic we have

$$\text{Rank}(|x_i|): 3, \ 1.5, \ 1.5$$

and using the tabulation of signs above, the following distribution is obtained

$$w: \quad -6 \quad -3 \quad 0 \quad 3 \quad 6$$

$$P(W = w): \quad \frac{1}{8} \quad \frac{1}{4} \quad \frac{1}{4} \quad \frac{1}{4} \quad \frac{1}{8}$$

The variance of this distribution is $1.5^2 + 1.5^2 + 3^2 = 13.5$; in the case of no ties with $n = 3$, $\text{var}(W) = 1^2 + 2^2 + 3^2 = 14$.

2.4 Asymmetric distributions: M-estimates

For asymmetric distributions the median is perhaps the most natural measure of location. However, as we have indicated in Section 2.1, a measure of location for an arbitrary $F(x)$ can be defined as the solution of the following equation in t:

$$\int_{-\infty}^{\infty} \psi(x - t)\, dF(x) = 0 \qquad (2.58)$$

where $\psi(x - t)$ is some suitably chosen function. Typically $\psi(u)$ should be monotonic in u, and may be bounded, thus ensuring regular behaviour of the statistics associated with the estimating equation corresponding to (2.58). The estimating equation is

$$M_\psi(\mathbf{X}, t) = (1/n) \sum_{i=1}^{n} \psi(X_i - t) = 0 \qquad (2.59)$$

Since the arguments based on symmetry are not applicable, it appears that only approximate inference about the location parameter, θ, is possible, based on large-sample approximate normality of the distribution of M_ψ and estimation of its variance. With suitably chosen $\psi(u)$ the normal approximation of M is reasonable since it is the sum of identically and independently distributed random variables. However, the effect of having to use an estimate of the variance of M_ψ is uncertain.

Even if the statistical problems of inference about θ could be overcome, the difficulty of interpreting the resulting estimate as a measure of location remains. In certain comparative studies, for example in the two-sample location problem, this problem of interpretation does not arise. In such cases the robustness properties of the transformations can be advantageous.

Exercises

2.1 The data given below are $n = 91$ gold assay results (from Krige, D.G. (1952) A statistical analysis of some of the borehole values in the Orange Free State Goldfields. *J. Chem. (Metall. Min.) Soc. S. Afr.*, **53**, 47)

Find a point estimate and two sided 95% confidence limits for the population median assay value.

1, 1, 5, 5, 5, 8, 11, 12, 12, 14, 14, 15, 24, 24, 33, 34, 37, 39, 39, 41, 43, 45, 47, 48, 51, 53, 53, 54, 54, 56, 59, 64, 79, 80, 83, 85, 88, 92, 94, 96, 104, 108, 109, 109, 129, 143, 149, 150, 157, 160, 166, 170, 180, 188, 191, 195, 198, 201, 210, 222, 227, 227, 238, 241, 244, 261, 310, 312, 327, 336, 349, 376, 383, 388, 400, 405, 421, 437, 439, 518, 546, 665, 678, 890, 906, 1009, 1085, 1747, 1893, 2898, 23036.

2.2 In sampling from an exponential population with median $\theta = 1$, calculate the expectation of the sample median when $n = 5$, thus verifying by example that the sample median is in general not an unbiased estimator θ.

2.3 Show that the sample median $\hat{\theta}$ is a median unbiased estimate of the population median θ.

2.4 Quantiles other than the median can be treated by the methods of Section 2.2. Consider the lower quantile ξ. An estimating equation for ξ is

$$C(\mathbf{x}, \xi) - n/4 = Q$$

where $C(\mathbf{x}, \xi)$ is the number of observations $x_1, x_2, \ldots, x_n$ smaller than ξ. The null distribution of $C(\mathbf{x}, \xi)$ is Binomial $(n, \frac{1}{4})$.

Using the data in Exercise 2.1, find a point estimate and two-sided 90% confidence limits for ξ.

2.5 Consider taking a random sample of size $n = 5$ from a $R(-\frac{1}{2} + \theta, \frac{1}{2} + \theta)$ population and testing $H_0 : \theta = 0$ against $H_1 : \theta > 0$. Tabulate the exact distribution of the Wilcoxon signed rank statistic W under the alternative $H_1 : \theta = \theta_1 = 0.05$.

Hint: note that under H_1 conditionally on $|X| < 0.45$,

$$\Pr\{\text{sgn}(X) = +1\} = \tfrac{1}{2}.$$

2.6 In Exercise 2.5 derive expressions in terms of n and θ_1 for

$$E(W | H_1) \text{ and } \text{var}(W | H_1).$$

2.7 Suppose that the distribution of X is uniform between $-\frac{1}{2} + \theta$ and $\frac{1}{2} + \theta$. Obtain an expression for the efficacy of W_G if $G(u) = u^\beta$, $\beta > 0$.

2.8 Obtain the efficacies of the sign statistic S and the Wilcoxon statistic W when the distribution of X has density $K\{1 + (x - \theta)^2\}^{-m}$ where K is a constant. Find the value of m for which the two statistics have the same efficacy.

2.9 The frequency distribution below is typical of distributions of errors of measurement of latitude; for convenience the units have been adjusted to give the class centres as shown.

Assuming the underlying distribution to be symmetric, obtain a two-sided 95% confidence interval for its centre using the Wilcoxon statistic W.

Class centre	Frequency	Class centre	Frequency	Class centre	Frequency
-8	1	-2	3	4	2
-7	0	-1	6	5	0
-6	0	0	10	6	1
-5	0	1	9	7	0
-4	1	2	7	8	0
-3	1	3	4	9	1

CHAPTER 3

Miscellaneous one-sample problems

3.1 Introduction

While one of the dominant themes of this book is distribution-free inference about location, this chapter gives attention to some traditional one-sample problems that have not been dealt with in Chapter 2. These have to do with dispersion and with various aspects of estimating the distribution function and of density estimation. One of the applications of density estimation is to estimation of the standard deviations of certain distribution-free estimators. Although such standard deviation estimates are not always needed for interval estimation, they can be useful in certain problems of combining estimates, in problems involving nuisance parameters, and serve to give an indication of the precision of estimates. Some attention will be given to the problem of estimating standard deviations of estimates. Estimation of the distribution function (or its complement) in the presence of censored observations is an important practical problem that is also discussed.

3.2 Dispersion: the interquartile range

The standard deviation or its square, the variance, is by far the most widely used measure of dispersion of a distribution (or of a random variable). Many reasons can be advanced for its popularity, among them that the variance can be interpreted as an expected squared error, and that when it exists for a location-scale parameter distribution, the standard deviation is proportional to the scale parameter. However, in the distribution-free context much the same reasons that make the mean unsuitable as a measure of location also apply to the standard deviation. Notably, there is its possible non-existence.

For a continuous distribution function $F(x)$ the interquartile range is

$$\Delta = \xi_{0.75} - \xi_{0.25}$$

where $\xi_{0.25}$ and $\xi_{0.75}$ are quartiles defined by

$$F(\xi_p) = p, \qquad p = 0.25, \, 0.75$$

The interquartile range is clearly a measure of dispersion of F, readily interpretable, and like the median, it always exists. For the well-known standard distributions like the normal and the Cauchy, it is easy to express the usual scale parameter in terms of Δ. For example, in the standard normal case $\Delta = 1.348$.

Despite its simple definition and interpretation, inference about Δ is not straightforward, in general. Essentially this comes about because, using a sample of x observations, Δ is estimated by the difference of sample quartiles whose joint distribution can be complicated. This matter will be touched upon again in Section 3.2.2, after considering a much simpler special case in Section 3.2.1.

3.2.1 *Symmetric F, known location*

We shall assume F to be symmetric about 0. Then, denoting the number of random observations $x_1, x_2, \ldots, x_n$ on X that lie between $-\Delta/2$ and $+\Delta/2$ by n_Δ, we note that the distribution of n_Δ is binomial $(n, \frac{1}{2})$. This fact can be used to test a hypothesis specifying a value of Δ, and to find a confidence interval for Δ.

Example 3.1 $n = 12$ observations from a symmetric distribution centred at 0:

$$-1.86, \, -1.01, \, -0.87, \, -0.65, \, -0.50, \, -0.12$$
$$-0.06, \quad 0.29, \quad 0.59, \quad 0.72, \quad 0.97, \quad 1.41$$

Testing $H_0 : \Delta = 2.0$, $H_1 : \Delta > 2.0$
H_0 will be rejected in favour of H_1 if observed n_Δ is sufficiently large.

$$\text{Observed } n_\Delta = 7$$
$$\Pr\{n_\Delta \geqslant 7 | H_0\} = 739/4096$$

thus H_0 is accepted at the conventional levels of significance.

Confidence limits for Δ
If the true Δ is used to find n_Δ we have

$$\Pr\{3 \leqslant n_\Delta \leqslant 10\} = 1 - 158/4096 = 0.9614$$

Now $\Delta > 2(1.86) = 3.72$ yields $n_\Delta = 12$

$$3.72 > \Delta > 2(1.41) = 2.82 \text{ yields } n_\Delta = 11$$

etc. Proceeding in this way, a two-sided 96% confidence interval for Δ is seen to be

$$(0.58, 2.82)$$

Point estimation

A point estimate $\bar{\Delta}$ of Δ is obtained by solving for Δ the equation

$$S(\mathbf{X}, \Delta) = n_\Delta - n/2 = 0$$

We observe that

$$E\{S(\mathbf{X}, d)\} = n\{F(d/2) - F(-d/2)\}$$

and $\quad (\partial E\{S(\mathbf{X}, d)\}/\partial d)_{d=\Delta} = nf(\Delta/2)$

Applying the results of Section 1.6.1, the large-sample variance of $\bar{\Delta}$ is approximately given as

$$\operatorname{var}(\bar{\Delta}) \simeq 1/\{4nf^2(\Delta/2)\}$$

As an indication of relative efficiency we can calculate the variance of the MLE of Δ in the case of a $N(0, \sigma^2)$ distribution. The MLE of Q is

$$\hat{\Delta} = 1.348 \left[\sum_{i=1}^{n} (x_i - \bar{x})^2/n \right]^{1/2}$$

and its variance for large n is given by

$$\operatorname{var}(\hat{\Delta}) \simeq (1.348)^2 \sigma^2/(2n)$$

while $\quad \operatorname{var}(\bar{\Delta}) \simeq \sigma^2/[4n\phi^2(0.6745)]$

giving $\quad \operatorname{var}(\hat{\Delta})/\operatorname{var}(\bar{\Delta}) \simeq 0.37.$

3.2.2 *General F*

Testing a hypothesis specifying a value of Δ is much more difficult in this case because, in order to examine the compatibility of a given value of Δ with the sample results, a decision has to be made about the location of the interquartile interval. Suppose that the left end-point of the interval is at ξ, then its right end-point is at $\xi + \Delta$, and suppose that the numbers of observations in the intervals thus generated are $N_1(\xi)$, $N_2(\xi)$, $N_3(\xi)$, with $N_1 + N_2 + N_3 = n$.

Then the log likelihood of the observed configuration

$$\log L(\xi, \Delta) = \log n! - \sum_{j=1}^{3} \log(N_j(\xi))!$$
$$- \{N_1(\xi) + N_3(\xi)\} \log 4 - N_2(\xi) \log 2.$$

and one possibility is to choose ξ so as to maximize $\log L(\xi, \Delta)$. One way of defining a test statistic is to base it on the difference $\max_{\xi} \log(\xi, \Delta) - \max_{\xi, \Delta} \log(\xi, \Delta)$, noting that for large n with, for convenience, n a multiple of 4,

$$\max_{\xi, \Delta} \log(\xi, \Delta) = \log n! - 2\log(n/4)! - \log(n/2)!$$
$$- (n/2)(\log 4 + \log 2)$$

In small samples there are several difficulties in this approach, one of them being that there is not necessarily a unique ξ which maximizes $\log(\xi, \Delta)$ for fixed Δ. More seriously, the null distribution of the proposed statistic is not distribution-free (it depends on F).

Similar difficulties attend interval and point estimation of Δ. The obvious point estimate of Δ is

$$D = \hat{\xi}_{0.75} - \hat{\xi}_{0.25}$$

where $\hat{\xi}_{0.75}$ and $\hat{\xi}_{0.25}$ are sample quartiles. The exact definition of $\xi_{0.25}$ (and $\xi_{0.75}$) is a matter of convention; one possible procedure is to smooth the sample distribution function $F_n(x)$ which has steps of size $1/n$ at the order statistics $x_{(1)}, x_{(2)}, \ldots, x_{(n)}$ as follows: The ordinate of $F_n(x)$ for $x_{(r)} < x < x_{(r+1)}$, is r/n, $r = 1, 2, \ldots, n-1$. Let the smoothed $F_n(x)$ have ordinates r/n at $\bar{x}_r = (x_{(r)} + x_{(r+1)})/2$ and interpolate linearly for x values between $\bar{x}_r$, and $\bar{x}_{r+1}$.

If n is sufficiently large to allow replacement without serious error of $\hat{\xi}_{0.25}$ and $\hat{\xi}_{0.75}$ by $x_{(r)}$ and $x_{(s)}$ respectively, where $r = [n/4]$, $s = [3n/4]$, we have from standard results about order statistics:

$$E(D) \simeq \Delta$$

$$\text{var}(D) \simeq [3f(\xi_{0.75}) + 2f(\xi_{0.25})f(\xi_{0.75})$$
$$+ 3f^2(\xi_{0.25})]/[16nf^2(\xi_{0.25})f^2(\xi_{0.75})] \qquad (3.1)$$

The discussion above, and especially the form of the expressions in (3.1), raises a problem which it has been possible to avoid in the one-sample, one-parameter problem of Chapter 2, namely, taking account of a nuisance parameter. In this case the nuisance parameter is the left location of the interquartile interval, or $\xi_{0.25}$ in (3.1), where $\xi_{0.75} = \xi_{0.25} + \Delta$. Generally in cases of this sort it is impossible to devise exact inferential procedures. With large n one can proceed by estimating $\text{var}(D)$.

Estimates of quantities such as $\text{var}(D)$, as given by (3.1), and of densities such as $f(\xi_{0.75})$, or the density at the median of X, depend fundamentally on estimating the distribution function $F(x)$. In the sections that follow we review some established results concerning inference about $F(x)$. We shall return, later, to the question of density estimation.

3.3 The sample distribution function and related inference

The sample distribution function, $F_n(x)$, is defined as

$$F_n(x) = (1/n) \quad \text{(the number of observations } x_1, x_2, \ldots, x_n \leqslant x)$$

and in Chapter 2 its fundamental role in the construction of estimators was illustrated. Generally we are interested in parameters that are defined by certain operations involving the distribution function F. If F_n is substituted for F in such an operation the result is, usually, that a consistent estimator of the parameter is obtained.

Thus it is important to note that, at every value of x, $F_n(x)$ is a consistent estimator of $F(x)$. It is, in fact the maximum-likelihood estimator, and is unbiased with minimum variance, as is well known.

Apart from its uses indicated above, the sample distribution function is used directly in many tests of goodness of fit. In such tests a specific parametric F is often nominated and the goodness-of-fit test is really a check of the agreement between F_n and F. This is usually made in terms of a measure of distance between two distributions. With a completely specified F, such a test can always be cast as a test of goodness of fit of the observations $F(x_1)$, $F(x_2), \ldots, F(x_n)$ to a uniform $(0, 1)$ distribution. The test thus becomes distribution-free.

While the distribution-free property of such goodness-of-fit tests is comforting, we are, of course, speaking of tests of parametric hypotheses that happen to be distribution-free. When contemplating using the distribution-free techniques that form the major substance of this book we are not, generally, concerned with such tests. Therefore we shall look only at one of these tests adapted to the setting of confidence bands for F.

3.3.1 *One-sided confidence·bands for F*

Consider the function

$$F_n^*(x, d) = \min \{F_n(x) + d, 1\}$$

where $d > 0$. Obviously F_n^* lies entirely above F_n (except when their ordinates coincide) and it is possible that F_n^* also lies entirely above F.

Let us suppose that we can calculate the probability

$$\alpha(n, d) = \Pr\{F_n^*(x, d) \geqslant F(x) \text{ for every } x\}$$

Then $F_n^*(x, d)$, $x = (-\infty, \infty)$, describes a 100% one-sided confidence band for $F(x)$. The term 'confidence band' emphasizes that F_n^* is above F for every x. Confidence limits for F at selected x-values can, of course, be obtained in the standard way, using the binomial distribution.

Now, $F_n^*(x, d)$ will be entirely above $F(x)$ if

$$\left. \begin{array}{l} d > F(X_{(1)}) = U_1 \\ 1/n + d > F(X_{(2)}) = U_2 \\ d + 2/n > F(X_{(3)}) = U_3 \\ d + \rho/n > F(X_{(\rho)}) = U_\rho \end{array} \right\} \tag{3.2}$$

where ρ is an integer such that $d + \rho/n < 1, d + (\rho + 1)/n \geqslant 1$.

The joint distribution of $U_1, U_2, \ldots, U_n$ is known its p.d.f. being

$$n! \text{ for } 0 \leqslant u_1 \leqslant u_2 \leqslant \ldots \leqslant 1$$

Hence the probability of the event described by (3.2), that is, $\alpha(n, d)$, can be computed by evaluating an appropriate integral:

$$\alpha(n, d) = n! \int\limits_{u_1 = 0}^{d} \int\limits_{u_2 = u_1}^{d + 1/n} \cdots \int\limits_{u_\rho = u_{\rho - 1}}^{d + \rho/n} \int\limits_{u_{\rho + 1} = u_\rho}^{1}$$

$$\int\limits_{u_n = u_{n-1}}^{1} du_1 \ldots du_n \tag{3.3}$$

The form of the integral in (3.3) in which F does not appear emphasizes the distribution-free character of the confidence band.

Straightforward calculations give

$$\alpha(n, a) = n! \int\limits_{u_1 = 0}^{d} \int\limits_{u_{\rho + 1} = u}^{d + \rho/n} \frac{(1 - u_{\rho + 1})^{n - \rho - 1}}{(n - \rho - 1)!} du_{\rho + 1} \ldots du_1 = n! G_n(\rho),$$

and

$$G_n(\rho) = G_n(\rho - 1) - \frac{(1 - \alpha - \rho/n)^{n - \rho - 2}}{(n - \rho - 2)!} H_n(\rho - 1)$$

Therefore, by successive addition we can write

$$G_n(\rho) = G_n(0) - \sum_{i = 1}^{\rho} \frac{(1 - d - i/n)^{n - i}}{(n - i)!} H_n(i - 1).$$

Further we note,

$$H_n(0) = d$$

$$H_n(1) = \frac{d}{2}(d + 2/n)$$

$$H_n(r) = \frac{d}{(r+1)!}\left(d + \frac{r+1}{n}\right)$$

so that

$$\alpha(n, d) = n! G_n(0) - d \sum_{i=1}^{\rho} \binom{n}{i}(1 - d - i/n)^{n-i}(d + i/n)^{i-1}$$

(3.4)

with $G_n(0) = 1 - (1 - d)^n$.

An approximation to $\alpha(n, d)$ for large n can be obtained by putting $d = \lambda/\sqrt{n}$, using Stirling's approximation for $n!$ which leads to the sum in (3.4) being approximated by an integral of the form

$$\frac{1}{\sqrt{2\pi}} \int_0^1 \frac{1}{\sqrt{y(1-y)}} \exp[-\lambda^2/(2y(1-y))]\, dy$$

By a substitution $y = \frac{1}{2} + u$ followed by $4u^2/(1 - 4u^2) = z^2$, this integral can be shown to be approximately $(1/\lambda)\exp(-2\lambda^2)$. Since a is taken to be finite, large n implies large λ.

Substitution in (3.4) finally gives the large-sample approximation

$$\alpha(n, d) = 1 - \exp(-2\lambda^2).$$

The derivation given above is in Wilks (1962, p. 336).

Two-sided confidence bands can be set in a manner similar to that discussed above. Calculation of the confidence coefficient is, however, considerably more complicated; the reader may refer to one of many texts that deal with the problem, or again to Wilks (1962, p. 339).

3.3.2 *Estimation of densities, and some related topics*

Estimation of densities arises in a variety of statistical problems; in the context of distribution-free techniques it turns out, as we have already noted in Chapter 2, that the density of the underlying F appears in expressions for asymptotic variances and relative efficiencies. Thus, if one were to estimate the variance of the sample median, one way of proceeding would be to use the asymptotic formula for the variance of the sample median, and attempt to estimate $f(\theta)$.

Since $F_n(x)$ is a consistent estimate of $F(x)$ for every x, it seems reasonable to suppose that one might take as an estimate of $f(x)$,

$$f_n(x) = [F_n(x + h) - F_n(x - h)]/(2h)$$

with h suitably chosen. In order that $f_n(x)$ be a consistent estimator of $f(x)$ it is clear that h should be chosen as a function, $h(n)$, of n, such that $h(n) \to 0$ as $n \to \infty$. The question of the rate at which $h(n)$ should approach 0 arises, and to answer it the statistical properties of $f_n(x)$ have to be examined.

Following Parzen (1962) the estimate $f_n(x)$ can be put in the form

$$f_n(x) = \frac{1}{h} \int_{-\infty}^{\infty} K\left(\frac{x-y}{h}\right) dF_n(x) = \frac{1}{nh} \sum_{j=1}^{n} K\left(\frac{x-x_j}{h}\right) \quad (3.5)$$

where

$$K(y) = \begin{cases} \frac{1}{2} & |y| \leqslant 1 \\ 0 & |y| > 1 \end{cases}$$

Such a representation has two advantages, the first being that the form of (3.5) suggests that K need not be restricted to the simple choice given above. A variety of functions are suitable for K and some of these are shown in Table 3.1; this table is extracted from Parzen (1962). The potential gain in having a wider choice of K is that some functions K may have a greater smoothing effect, leading to more reliable estimates.

Table 3.1

$K(y)$	$\int_{-\infty}^{\infty} K^2(y) dy$
$\begin{cases} \frac{1}{2} & |y| \geqslant 1 \\ 0 & |y| > 1 \end{cases}$	$\frac{1}{2}$
$\frac{1}{2}\exp(-|y|)$	$\frac{1}{2}$
$(1/2\pi)[\sin(y/2)/(y/2)]^2$	$1/3\pi$

The other advantage of writing $f_n(x)$ as in (3.5) is that it is seen to be the sum of independent identically distributed random variables, thus facilitating the investigation of its statistical properties. In this investigation the following theorem is used.

Theorem 3.1 Suppose that $K(y)$ satisfies:

(i) $\sup_{-\infty < y < \infty} |K(y)| < \infty$

(ii) $\displaystyle\int_{-\infty}^{\infty} |K(y)|\, \mathrm{d}y < \infty$

(iii) $\displaystyle\lim_{y \to \infty} |y K(y)| = 0$

Also suppose that $g(y)$ satisfies

$$\int_{-\infty}^{\infty} |g(y)|\, \mathrm{d}y < \infty$$

and that $h(n) \to 0$ as $n \to \infty$. Then at every point of continuity of $g(y)$,

$$\lim_{n \to \infty} g_n(x) = \lim_{n \to \infty} \frac{1}{h(n)} \int_{-\infty}^{\infty} K\left(\frac{y}{h(n)}\right) g(x - y)\, \mathrm{d}y$$

$$= g(x) \int_{-\infty}^{\infty} K(y)\, \mathrm{d}y$$

The proof of this theorem is obtained by noting that

$$\Delta_n(x) = g_n(x) - g(x) \int_{-\infty}^{\infty} K(y)\, \mathrm{d}y$$

$$= \int_{-\infty}^{\infty} [g(x - y) - g(x)] \frac{1}{h(n)} K\left(\frac{y}{h(n)}\right) \mathrm{d}y,$$

and splitting the region of integration into two regions $|y| \leqslant \delta, \delta > 0$, and $|y| > \delta$, and considering $|\Delta_n(x)|$. The upper bound to $|\Delta_n(x)|$ comprises contributions which tend to 0 as first $n \to \infty$ and then $\delta \to 0$.

Consider the expectation of $f_n(x)$ as $n \to \infty$. From (3.5) we see that

$$E\{f_n(x)\} = E\left\{\frac{1}{h(n)} K\left(\frac{x - X}{h(n)}\right)\right\} = \int_{-\infty}^{\infty} \frac{1}{h(n)} K\left(\frac{x - y}{h(n)}\right) f(y)\, \mathrm{d}y$$

Applying Theorem 3.1, we see that

$$\lim_{n \to \infty} E\{f_n(x)\} = f(x)$$

if the conditions stated in the theorem are satisfied and if $K(y)$ is scaled so that

$$\int_{-\infty}^{\infty} K(y)\,dy = 1$$

We assume this to hold in further discussions.

Thus $f_n(x)$ is asymptotically unbiased at every point of continuity of $f(x)$, and in order to establish conditions for its consistency we need only examine var $[f_n(x)]$. Again using (3.5),

$$\text{var}\,[f_n(x)] = (1/n)\text{var}\,[(1/h)K((x-X)/h)]$$

$$= (1/n)\left\{ \frac{1}{h^2(n)} \int_{-\infty}^{\infty} K^2\left(\frac{x-y}{h(n)}\right) f(y)\,dy \right.$$

$$\left. \left[\int_{-\infty}^{\infty} \frac{1}{h(n)} K\left(\frac{x-y}{h(n)}\right) dy \right]^2 \right\}$$

so that

$$nh(n)\text{var}\,[f_n(x)] = \int_{-\infty}^{\infty} \frac{1}{h(n)} K^2\left(\frac{x-y}{h(n)}\right) f(y)\,dy$$

$$- h(n)\left[\int_{-\infty}^{\infty} \frac{1}{h(n)} K\left(\frac{x-y}{n}\right) f(y)\,dy \right]^2 \tag{3.6}$$

Using Theorem 3.1 again, we see that

$$\lim_{n \to \infty} nh(n)\text{var}\,[f_n(x)] = f(x) \int_{-\infty}^{\infty} K^2(y)\,dy \tag{3.7}$$

The result (3.7) shows that $f_n(x)$ will be mean-square consistent if in addition to

$$\lim_{n \to \infty} h(n) = 0$$

we also have

$$\lim_{n \to \infty} nh(n) = \infty$$

These results indicate that $h(n) = wn^{-1/2}$, w constant, might be an appropriate choice of $h(n)$ for mean-square consistency. However, the choice of w in any particular example is not obvious. Rosenblatt (1971) outlines an argument showing that the rate at which the mean square error of estimation tends to 0 is maximized if

$$h(n) = wn^{-1/5}$$

a suitable value of w being

$$w = \{4f(x) \int K^2(y) \, dy / [f''(x) \int g^2 K^2(y) \, dy]\}^{1/5}$$

Except as a rough guide this formula is not of much practical use as it involves $f''(x)$; it could possibly be used in an iterative manner, an initial estimate of $f(x)$ (and $f''(x)$) being used in the place of $f'(x)$ (and $f''(x)$).

Example 3.2 Refer to the $n = 91$ sample assay values of Exercise 2.1. We shall perform only some rather crude calculations to illustrate procedures.

Grouping the observations into classes of width 50 units we obtain the following frequency distribution; class lower limits only are shown.

Class	0–	50–	100–	150–	200–	250–	300–	350–	400–	450–	500–
Frequency	24	16	7	10	8	1	5	3	5	0	12
K		0.020	0.032	0.039	0.038	0.028	0.017	0.007	0.003	0.001	0.00

The observed sample median is 143 and this falls in the class with lower limit 100. A crude estimate of the density at the mid-point of this interval is $(7/91)(1/50) = 0.001\,54$. One would expect a more refined estimate of the density at $x = 143$ to be of the same order of magnitude.

Taking $K(y) = (\sigma \sqrt{2})^{-1} \exp(-y^2/(2\sigma^2))$ with $\sigma = 10$ and $h = 10$, the values of $K[(143 - x_i)/10]$ at the class mid-point values x_i are shown in the table above. The mean K-value is 0.0212 giving $f_n(x) = 0.002\,12$.

One way in which this result might be used is in setting approximate confidence limits for the median using approximate normality of the distribution of the sample median. The estimated standard deviation of the sample median is then

$(4 \times 91)^{-1/2}/(0.002\,12) = 24.7$, giving approximate two-sided confidence limits

$$143 \pm 1.645 \times 24.7 \simeq 143 \pm 41$$

These should be compared with the exact limits obtained as in Exercise 2.1.

The ideas outlined above for estimating $f(x)$ can be applied to other cases where an estimate is required of the slope of a continuous function, which is itself the expectation of a sample function that is not necessarily continuous. An example of this kind was discussed in Chapter 2, namely the Wilcoxon statistic $W(\mathbf{x}, t)$ defined in equation (2.23).

It will be recalled that $W(\mathbf{X}, t)$ can be put in the form

$$W^*(\mathbf{X}, t) = n(n+1)/2 - 2 \# \left(\frac{x_i + x_j}{2} \leqslant t \right)$$

Now the expectation of the second term on the r.h.s. of the above expression is

$$\frac{n(n+1)}{2} F_2(t)$$

where $F_2(t)$ is the distribution function of $(X_i + X_j)/2$, X_i and X_j being independent and distributed like X. Putting

$$F_{2,n}(t) = \frac{2}{n(n+1)} \# \left(\frac{X_i + X_j}{2} \leqslant t \right)$$

and

$$f_{2,n}(t) = \frac{2}{n(n+1)h(n)} \sum_{i,j=1}^{n} K\left(\frac{t - (x_i + x_j)/2}{h(n)} \right)$$

where $K(y)$ is defined as above, the arguments used before show that $f_{2,n}(t)$ is asymptotically unbiased for $f_2(t) = F_2'(t)$.

It is possible to establish conditions for $h(n)$ to ensure that $f_{2,n}(t)$ is consistent for $f_2(t)$. The calculations are complicated and details will not be given here.

3.3.3 Direct estimates of variances of estimates

Consider estimation of the standard deviation of the sample median $(\hat{\theta})$. The asymptotic formula for var $(\hat{\theta})$ involves $f(\hat{\theta})$, the density at the true median value. It can be estimated by one of the methods outlined in Section 3.2, and the estimated value used in the asymptotic formula. However, var $(\hat{\theta})$ can be estimated more directly for we have, in the case $n = 2m + 1$,

$$E(\hat{\theta}^x) = E(X_{(m+1)}) = \frac{n!}{(m!)^2} \int u^r F_{(u)}^m (1 - F_{(u)})^m f(u) \, du$$

$$= \frac{n!}{(m!)^2} \int_0^1 \psi^r(y) y^m (1 - y)^m \, dy$$

where $\psi(y) = F^{-1}(y)$. Now the sample c.d.f. provides an estimate of $F(y)$, and hence of $\psi(y)$, thus leading to an explicit estimate of the actual variance of $\hat{\theta}$. This method is discussed more fully in Maritz and Jarrett (1978).

Similar procedures can be applied to any estimates that are linear functions of order statistics, for example, the interquartile range.

3.4 Estimation of F when some observations are censored

We shall consider only non-negative random variables and the case of right censoring of some of the observations, by which is meant the following: an observation is said to be *right censored* at the value x if it is known that its actual value is $\geq x$. Such censored observations arise typically in studies of survival, where patients may be lost to the study for reasons unconnected with the agencies that may affect their survival. For example, a patient may be transferred to another working place by his employer. In studies of this type it is common practice to focus attention on the complement of $F(x)$, the *survivorship function* $F_s(x) = 1 - F(x)$.

3.4.1 *The actuarial method of estimating F*

When the number of observations is large, the data are conveniently summarized in the form of a life table with some grouping of observations, as illustrated in Table 3.2; for obvious reasons this procedure for estimating F is called an 'actuarial method' (Pike and Roe, 1963). The classes need not be of equal length and we denote them $x_0 - x_1, x_1 - x_2$, etc., with $x_0 = 0$. The steps in calculating an estimate of $F_s(x)$ are shown in the same table, resulting in the entries in column (7) which are estimated values of F_s at the lower end-points of the successive class intervals.

Let the total number of patients be n. It is convenient to think of all

Table 3.2

(1) Interval	(2) Deaths	(3) With- drawls	(4) Number at risk at x_r	(5) n_r	(6) p_r	(7) $\hat{F}_s(x_r)$
$x_0 - x_1$	d_0	w_0	n_0	$n_0 - w_0/2$	$p_0 = d_0/n_0'$	1
$x_1 - x_2$	d_1	w_1	n_1	$n_1 - w_2/2$	$p_1 = d_1/n_1'$	p_0
$x_2 - x_3$	d_2	w_2	n_2	$n_2 - w_2/2$	$p_2 = d_2/n_2'$	$p_0 p_1$

patients as having been first seen at time 0 and then followed until death or withdrawal through other causes. The total number is also the number at risk at the start of the first x-interval, and we denote this number $n_0 = n$. The number of deaths and withdrawals during this first interval are, respectively, d_0 and w_0. The number at risk at the start of the next interval is $n_0 - d_1 - w_1$. The rest of the entries in columns (1)–(4) are interpreted similarly.

If there were no withdrawals, the ratio d_r/n_r would be an estimate of the conditional probability of dying in the rth interval, given survival up to the start of the interval. Since a number w_r of patients withdraw during the interval, the effective number at risk for the interval is between n_r and $n_r - w_r$, and is usually taken to be $n_r - w_r/2 = n_r'$. The conditional probability of dying is therefore estimated by

$$q_r = d_r/(n_r - w_r/2)$$

Denoting the estimate of $F_s(x)$ by $\hat{F}_s(x)$, we then have

$$\hat{F}_s(x_{r+1}) = \hat{F}_s(x_r) \cdot (1 - q_r) = \hat{F}_s(x_r)p_r$$

Since $\hat{F}_s(x_0) = 1$, we have

$$\hat{F}_s(x_1) = p_0$$
$$\hat{F}_s(x_2) = p_0 p_1$$

etc.

Example 3.3 Table 3.3 gives a numerical illustration of the estimation procedure outlined above; it is taken from Armitage (1971, p. 412); $n = n_0 = 374$.

A large-sample formula for var $F_s(x_r)$ can be developed as follows. Write $F_s(x_r) = \exp(L_r)$, where

$$L_r = \ln(p_0 p_1 \cdots p_{r-1})$$

Table 3.3

(1) Interval	(2) Deaths	(3) Withdrawls	(4) n_r	(5) n_r'	(6) p_r	(7) $\hat{F}_s(x_r)$
0–1	90	0	374	374.0	0.7594	1.000
1–2	76	0	284	284.0	0.7324	0.759
2–3	51	0	208	208.0	0.7548	0.556
3–4	25	12	157	151.0	0.8344	0.420
4–5	20	5	120	117.5	0.8298	0.350
5–6	7	9	95	90.5	0.9227	0.291
6–7	4	9	79	74.5	0.9463	0.268
7–8	1	3	66	64.5	0.9845	0.254
8–9	3	5	62	59.5	0.9496	0.250
9–10	2	5	54	51.5	0.9612	0.237
10–	21	26	47			0.228

Now

$$\operatorname{var}\{F_s(x_r)\} \simeq \{F_s(x_r)\}^2 \operatorname{var}(L_r) \tag{3.8}$$

and we obtain an approximation for $\operatorname{var}(L_r)$ by noting, first, that

$$\operatorname{var}\{L_r \mid d_0, n_0', \ldots d_{r-2}, n_{r-2}', n_{r-1}'\} \simeq \left(\frac{1}{p_{r-1}}\right)^2 \frac{p_{r-1}(1 - p_{r-1})}{n_{r-1}'}.$$

Here we use again an approximation of the form $\operatorname{var}(\ln Y) \simeq \{1/E(Y)\}^2 \operatorname{var}(Y)$, and an assumption that d_{r-1} is binomially distributed. The latter assumption is strictly valid if $w_{r-1} = 0$.

Note also that

$$E\{L_r \mid d_0, n_0', \ldots, d_{r-2}, n_{r-2}', n_{r-1}'\}$$

$$= \ln\left(\frac{d_0}{n_0'}\right) + \ldots + k\left(\frac{d_{r-2}}{n_{r-2}'}\right) + \ln p_{r-1}$$

By repeatedly applying the formula

$$\operatorname{var}(Y) = E \operatorname{var}(Y \mid X) + \operatorname{var} E(T \mid X)$$

and successively reducing the number of conditioning variables we obtain Greenwood's (1926) formula

$$\text{var}(\hat{F}_s(x_r)) \simeq \{\hat{F}_s(x_r)\}^2 \sum_{i=0}^{r-1} \frac{d_i}{n_i'(n_i'-d_i)} \qquad (3.9)$$

When there are no withdrawals, Greenwood's formula simplifies to

$$\text{var}\{F_s(x_r)\} \simeq \hat{F}_s(x_r)\{1 - \hat{F}_s(x_r)\}/n_0$$

which is the elementary formula for the variance of a binomial random variable, with the exact probability replaced by its estimate.

One of the uses of Greenwood's formula is for testing a hypothesis that specifies a value for the median of X, and by the usual argument, it can then also be used to obtain a confidence interval for the median of X.

Example 3.4 Refer to Example 3.3 and test the hypothesis that $\theta = \text{median}(X) = 2.0$.

Using the version of Greenwood's formula with $\hat{F}_s(x_r)$ replaced by the theoretical value, $\frac{1}{2}$, that is, essentially using (3.8), we have

$$\text{var}\{F_s(2.0)\} \simeq (\tfrac{1}{4})\left\{\frac{90}{374(284)} + \frac{76}{284(208)}\right\} = (0.0231)^2.$$

Treating $\hat{F}_s(2.0)$ as normally distributed with the standard deviation calculated above, for a two-sided test at the 10% level we note that

$$0.5 \pm 1.645(0.0231) = 0.5 \pm 0.038.$$

Thus the null hypothesis stated above is rejected.

In order to determine a confidence interval for θ, the calculation performed above should be done at a succession of x-values. With the grouped data it is convenient, without regrouping, to perform the calculations at the class end-points. Interpolation using a graph as illustrated in Fig. 3.1 can be used to find a confidence interval. It will be noted that we plot graphs of

$$0.5 \pm 1.645(\text{s.d.}\,(\hat{F}))$$

and

$$\text{observed } \hat{F}_s(x_r)$$

against x. The x-values of appropriate points of intersection of these graphs give the confidence limits. The answers obtained from Figure 3.1 give a 90% confidence interval for θ as:

$$(2.14, 2.76)$$

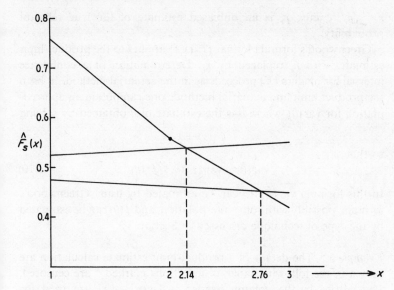

Figure 3.1 *Graphical interpolation to obtain confidence limits for the median from grouped censored data.*

3.4.2 *The product-limit method of estimating F*

The product-limit method can be regarded as a refinement of the actuarial method, effected by a judicious choice of class boundaries. Let $\xi_1 < \xi_2 < \xi_3 < \ldots < \xi_n$ be the observed lifetimes, censored or uncensored, and suppose that the class boundaries are $\xi_1 +$, $\xi_2 +, \ldots, \xi_n +$. Then, in the notation of Section 3.4.1, every d is either 0 or 1, and every w_r is either 1 or 0. Also, when an interval does contain a loss, that is, $w_r = 1$, the loss occurs just before the end of the interval so that n_r need not be adjusted to n'_r as was done in the actuarial method. Thus

$$p_r = \begin{cases} 1 - 1/n_r & \text{if death at } \xi_{r+1} \\ 1 & \text{if loss at } \xi_{r+1} \end{cases}$$

and we have, as before

$$\hat{F}_s(x_r) = p_0 p_1 \ldots p_{r-1}$$

The product-limit method is described by Kaplan and Meier (1958) in a paper that also contains much other useful material about the analysis of censored data. Among these results is a demonstration that the product-limit estimate is actually a maximum-likelihood estimate of $F_s(x)$. Also, $\hat{F}_s(x_r)$ is unbiased for $F_s(x)$, as can be seen by calculating its expectation by successive conditioning up to the term

p_{r-2}, p_{r-3}, etc.; p_r is an unbiased estimate of the true relevant probability.

Greenwood's formula for var $\{\hat{F}_2(x_r)\}$ applies to the product-limit estimate, with n_r' replaced by n_r. Determination of a confidence interval for median (X) proceeds as in the actuarial method. In both the product limit and actuarial methods one can obtain an approximation for var $(\hat{\theta})$, where $\hat{\theta}$ is the estimate of θ obtained by solving

$$F_s(x) = \tfrac{1}{2}$$

so that

$$\text{var}(\hat{\theta}) \simeq \text{var}\{\hat{F}_s(\theta)\}/f^2(\theta) \qquad (3.10)$$

In this formula var $(\hat{F}_s(\hat{\theta}))$ can be estimated by using Greenwood's formula, possibly with some interpolation, and $f(\theta)$ can be estimated by the type of technique discussed in Section 3.3.

Example 3.5 The details of a product-limit estimate calculation are shown in the following table; observations marked * are censored. The entries in the column headed $d_r/[n_r(n_r - d_r)]$ are used for calculating estimates of var $\{\hat{F}_s(x_r)\}$.

ξ	n_r	d_r	p_r	$\hat{F}_s(r)$	$d_r/[n_r(n_r - d_r)]$
0.33	20	1	0.9500	0.9500	0.002 63
1.02	19	1	0.9474	0.9000	0.002 92
1.39*	18	1	0.9444	0.8500	0.003 27
1.47*	17	0	1.0000	0.8500	0
1.48*	16	0	1.0000	0.8500	0
1.67	15	1	0.9333	0.7933	0.004 76
1.84	14	1	0.9286	0.7366	0.005 49
2.09	13	1	0.9231	0.6800	0.006 41
2.15	12	1	0.9167	0.6234	0.007 58
2.51	11	1	0.9091	0.5667	0.009 09
2.70	10	1	0.9000	0.5100	0.011 11
4.08*	9	0	1.0000	0.5100	0
4.57*	8	0	1.0000	0.5100	0
4.60	7	1	0.8571	0.4371	0.023 81
4.88	6	1	0.8333	0.3643	0.033 33
4.98*	5	0	1.0000	0.3643	0
5.62*	4	0	1.0000	0.3643	0
8.11*	3	0	1.0000	0.3643	0
15.49	2	1	0.5000	0.1821	0.500 00
19.62	1	1	0.0000	0.0000	

From the table, $\hat{\theta}$ lies between 4.57 and 4.60; we take it at 4.585 for this illustration. Following the method of Example 3.2 with $h = 0.5$, $\sigma = 2.0$ the estimated value of $f(4.585)$ is 0.1155. Using Greenwood's formula the estimated values of var $\{\hat{F}_s(x_r)\}$ at 4.57 and 4.60 are 0.0138 and 0.0147. Substituting the average of these two values, and the estimated $f(4.585)$ in formula (3.10) the estimated standard deviation of $\hat{\theta}$ is 1.02.

CHAPTER 4

Two-sample problems

4.1 Types of two-sample problems

We shall suppose that our data comprise m independent observations, $x_1, x_2, \ldots, x_m$, on a random variable X with distribution function $F(x)$ and n independent observations, $y_1, y_2, \ldots, y_n$, on Y with distribution function $G(y)$. When necessary the dependence of these distribution functions on a parameter will be indicated. For convenience, the two samples may also be referred to as the X-sample or the Y-sample. Where convenient, the observations x_i will be regarded as realizations of independent identically distributed random variables X_i, $i = 1, 2, \ldots, m$, with common distribution function $F(x)$; a similar notation will be applied to the Y-sample.

In many applications of statistics, two-sample problems arise in such a way as to lead naturally to the formulation of a null hypothesis to the effect that the X- and Y-samples come from identical populations. For example, if subjects are randomly assigned to two groups in a trial to compare two treatments of hay fever, the responses being measured on some quantitative scale, one might begin with the null hypothesis that the treatments are equally effective.

The test procedure should be influenced by the alternative to the null hypothesis that is being contemplated. In this respect two-sample problems seem to be more complicated than one-sample problems; one has to describe the alternative hypothesis in terms of some reasonably easily interpreted measure of 'difference' between two distributions. Clearly a great diversity of measures of difference between two distributions can be defined.

Among such measures of difference, one of the simplest and most easily interpreted is a difference in location of distributions that are otherwise identical. In terms of the distribution functions F and G introduced above, this case of location difference is described by stating that $F(x) = F(x - \theta_1)$, $G(y) = F(y - \theta_2)$, the difference in location being $\theta_1 - \theta_2 = \Delta$. The main emphasis of this chapter will be on problems of location difference.

A somewhat more general measure of difference is the difference of medians of two distributions. This difference can be estimated without the restriction that F and G have the same 'shape'. In fact any quantile could be used in a similar manner. However, without the restriction of equal shape, equality of medians leaves room for two distributions to be quite different in other vital respects; this should be remembered in basing inference on comparison of medians only.

Another simple measure of difference between F and G which arises naturally in connection with rank tests is $\Pr(X < Y) - \frac{1}{2}$; its value is 0 when F and G are identical. An obvious estimate of $\Pr(X < Y)$ is the observed proportion of x_i-values smaller than y_i-values, and this is the well known Mann–Whitney statistic about which more details will be given later.

For the most part, this chapter will concentrate on test and estimation procedures that are associated with the measures of difference introduced above. Other measures and test procedures are also mathematically and practically important, but the attraction of the simpler measures is that they not only lead to useful test procedures, but that the associated point estimates are readily interpreted for practical use.

4.2 The basic randomization argument

Suppose that the null hypothesis H_0 holds, namely that F and G are identical. Then the X- and Y-samples can be regarded as having been drawn from the same population. Let us label the members

$$x_1, x_2, \ldots, x_m; \quad y_1, y_2, \ldots, y_n$$

of the pooled sample

$$z_1, z_2, \ldots, z_m, z_{m+1}, \ldots, z_N$$

where $N = m + n$. Then, under H_0, conditionally on the realization of the values $z_1, \ldots, z_N$, the particular observed X-sample can be regarded as having been obtained by a random selection without replacement of m of the z-values, labelled as $x_1, \ldots, x_m$. We shall also speak of a *random partition* of the z-values into groups of sizes m and n.

Under this scheme of random partitioning, the conditional null distribution of any proposed test statistic can, in principle, be obtained by listing all possible partitions and calculating the corresponding values of the statistic. The conditional null distribution generated in this way can then be used to perform a conditional test of significance and to set confidence limits. The arguments justifying such procedures are essentially the same as those used in Chapter 2. In the rest of this chapter the randomization argument is applied repeatedly in considering various test statistics.

4.3 Inference about location difference

4.3.1 *Introduction*

We now consider the case where the X- and Y-samples derive from populations that differ only in location. Thus $G(y) = F(y - \theta)$; in this formulation $\theta = E(Y) - E(X)$ when the expectations exist, and $\theta = \text{median}(Y) - \text{median}(X)$.

If the true location difference is θ the two sets of random variables

$$X_1, X_2, \ldots, X_m$$
$$Y_1 - \theta, Y_2 - \theta, \ldots, Y_n - \theta$$

are identically distributed. The randomization argument can therefore be applied to values $z_i(\theta)$, being the pooled collection of values $x_1, x_2, \ldots, x_m, y_1 - \theta, y_2 - \theta, \ldots, y_n - \theta$. We shall consider several statistics based on the $z_i(\theta)$ values.

4.3.2 *The two-sample mean statistic*

Let

$$A(\mathbf{X}, \mathbf{Y}, t) = \sum_{i=1}^{m} X_i - m\bar{z}(t), \qquad (4.1)$$

where $\bar{z}(t) = \sum_{i=1}^{N} z_i(t)/N$. We call this the *mean statistic* and it is, of course, inspired by the fact that the sample mean is a standard estimate of location. The statistic $A(\mathbf{X}, \mathbf{Y}, t)$ can also be expressed as

$$A(\mathbf{X}, \mathbf{Y}, t) = \frac{nm}{N}(\bar{X} - \bar{Y} + t) \qquad (4.2)$$

but since $\bar{z}(\theta)$ remains fixed in the randomization argument for finding the conditional null distribution of A, the form (4.1) is somewhat more convenient for most of our purposes. This statistic and its randomization distribution was studied in considerable detail by Pitman (1937).

The expression (4.2) shows that inference about θ using A amounts to the use of the difference of sample means, and we refer only briefly to the consistency and efficiency properties of the related procedures. If F has a finite variance σ^2, $A(\mathbf{X}, \mathbf{Y}, \theta)$ is asymptotically distributed $N\left(0, \sigma^2\left(\dfrac{m+n}{mn}\right)\right)$ as m and $n \to \infty$; the asymptotic distribution of $A(\mathbf{X}, \mathbf{Y}, t)$ is $N\left(t - \theta, \sigma^2\left(\dfrac{m+n}{mn}\right)\right)$. Consistency of testing and of estimation is readily established.

Testing $H_0 : \theta = \theta_0$ against $H_1 : \theta > \theta_0$

We concentrate on the distribution-free test procedure based on enumeration of the conditional null distribution of A. In practice we calculate the observed $A(\mathbf{x}, \mathbf{y}, \theta_0)$ and refer it to the null distribution. If the observed $A(\mathbf{x}, \mathbf{y}, \theta_0)$ is sufficiently small, that is, lies in the lower $100\alpha\%$ tail of this distribution, H_0 is rejected.

Example 4.1

$$x_i : 3.46, 4.13, 2.71$$
$$y_j : 4.85, 5.22, 5.64$$
$\theta_0 = 1.0$:
$$y_j - \theta_0 : 3.85, 4.22, 4.64$$
$$\bar{z}(\theta_0) = (3.46 + \ldots + 4.64)/6 = 3.835$$

Observed $A(\mathbf{x}, \mathbf{y}, \theta_0) = -1.205$

There are $\binom{6}{3} = 20$ ways of partitioning the $z_i(\theta)$ values into two groups of sizes 3 and 3 and the smallest values of A obtained are:

$$2.71 + 3.46 + 3.85 - 3(3.835) = -1.485$$
$$2.71 + 3.46 + 4.13 - 3(3.835) = -1.205$$
$$2.71 + 3.85 + 4.13 - 3(3.835) = -0.815$$

etc.

Thus $\Pr\{A(\mathbf{x}, \mathbf{y}, \theta_0) \leqslant -1.205\} = 2/20$ and we reject H_0 at the 10% level of significance.

Modification of the procedure illustrated above for two-sided alternatives is obvious.

Confidence limits for θ

Confidence limits for the location-shift parameter θ are set by the inversion of the hypothesis-testing procedure which we have used in earlier chapters. We shall consider a two-sided $100\left(1 - 2r \bigg/ \binom{m+n}{n}\right)\%$ confidence interval.

For every possible value t of θ we can calculate the observed $A(\mathbf{x}, \mathbf{y}, t)$ and by listing partitions of the $z_i(t)$ values we can find the number $N(t)$ of possible A-values that are greater than the observed A. Now, as t varies between $-\infty$ and $+\infty$, the value of $N(t)$ changes from $\binom{m+n}{n} - 1$ to 0, and to find the exact confidence limits we need to identify the t-values at which $N(t)$ changes.

In fact, $N(t)$ changes whenever

$$\sum_{i=1}^{m} x_i = \sum_{i=1}^{m} z_i^*(t) \tag{4.3}$$

where $z_1^*(t), z_2^*(t), \ldots, z_m^*(t)$ is a subset of all $z_i(t)$ values. If the r.h.s. of (4.3) contains $(m-1)$ of the x_i values and one $y_j - t$, then (4.3.) is satisfied by a t given by

$$x_i = y_j - t$$

Thus all values $y_j - x_i$ constitute change points of $N(t)$. Similarly, if the r.h.s. of (4.3) contains $(m-2)$ of the x_i values, the value of t satisfying (4.3) is given by

$$(x_i + x_j)/2 = (y_r + y_s)/2 - t$$

Thus the difference of averages $(y_r + y_s)/2 - (x_i + x_j/2)$ also constitute change points of $N(t)$.

To find the exact confidence limits all such differences of means of subsets of the same size could, in principle, be listed in order of magnitude. The confidence limits are, then, the rth smallest and rth largest of these differences. In practice it may be quicker to vary t from the largest value that needs considering, $y_{(n)} - x_{(1)}$, until the point at which $N(t)$ changes from $r - 1$ to r is found.

Example 4.2

$$x : 2.71, 3.46, 4.13$$

$$y : 4.85, 5.22, 5.64, 6.20$$

There are $4 \times 3 = 12$ differences of single observations

$$\binom{3}{2}\binom{4}{2} = 18 \text{ differences } (y_r - y_s)/2 - (x_i + x_j)/2$$

$$\binom{3}{3}\binom{4}{3} = 4 \text{ differences } (y_r + y_s + y_t)/3 - (x_i + x_j + x_k)/3$$

giving a total of $34 = \binom{7}{3} - 1$. Listed in decreasing order of magnitude, these differences are:

3.49	2.51	2.253	2.125	1.950	1.730	1.39
2.93	2.500	2.18	2.105	1.915	1.635	1.240
2.835	2.440	2.160	2.07	1.825	1.615	1.09
2.74	2.345	2.14	2.010	1.803	1.51	0.72
2.625	2.290	2.130	1.990	1.76	1.450	

Thus a $100(1 - 6/35)\% = 83\%$ confidence interval for θ is (1.240, 2.835).

Point estimation

The solution $\bar{\theta}$ of the estimating equation

$$A(\mathbf{X}, \mathbf{Y}, \theta) = 0$$

is $\bar{\theta} = \bar{Y} - \bar{X}$. This estimating equation is inspired by the fact that $E\{A(\mathbf{X}, \mathbf{Y}, \theta)\} = 0$ if it exists.

Large-sample approximations ·

Although it is not essential in hypothesis testing to list all partitions of the two samples, or in determining confidence limits to list all differences of subset means, exact calculations rapidly become prohibitive as m and n increase. Approximations of the conditional distribution, therefore, become important. The exact first two moments of the conditional distribution are

$$E(A) = 0$$

$$\text{var}(A) = \frac{mn\sigma^2(t)}{(m + n - 1)} \tag{4.4}$$

where $\sigma^2(t) = [1/(m + n)] \sum_{i=1}^{m+n} (z_i(t) - \bar{z}(t))^2$. These results follow from the standard theory of sampling at random without replacement from a finite population.

As m and $n \to \infty$ with $m/(m+n)$ approaching a limit λ, the distribution of $A/\text{s.d.}(A)$ approaches a standard normal distribution under conditions given in Section 1.8.3. Using this result the hypothesis-test procedure can be put in the form:

$$\text{reject } H_0 \text{ if } \bar{X} - \bar{Y} + \theta_0 < -u_\alpha \sigma(t) \sqrt{\frac{N^2}{(N-1)mn}}$$

The rule can be written in the form, reject H_0 if

$$\frac{\bar{X} - \bar{Y} + \theta_0}{s\left(\dfrac{1}{n} + \dfrac{1}{m}\right)^{1/2}} < -u_\alpha \left[\left(\frac{N-2}{N-1}\right)\frac{1}{\{1 - u_\alpha^2/(N-1)\}}\right]^{1/2} \quad (4.5)$$

where s denotes the usual 'within-group pooled standard deviation'. Thus the l.h.s. of (4.5) is the usual t-statistic, and in the event that F is a normal distribution, the r.h.s. of (4.5) can be regarded as giving an approximation to the appropriate quantile of the t-distribution. The following numerical values are instructive; the constant in the r.h.s. of (4.5) is denoted by k_α.

N	$\alpha = 0.9$		$\alpha = 0.95$	
	k_α	t_α	k_α	t_α
5	1.446	1.476	2.505	2.015
10	1.337	1.372	1.854	1.812
20	1.306	1.325	1.729	1.725

Using (4.5), the $100(1 - 2\alpha)\%$ confidence interval for θ is approximately

$$-\bar{X} + \bar{Y} \pm k_\alpha s[(m+n)/mn]^{1/2}$$

4.3.3 The two-sample sign statistic

Rewrite the mean statistic in (4.1) as

$$A(\mathbf{X}, \mathbf{Y}, t) = \sum_{i=1}^{m} [X_i - \bar{z}(t)] \quad (4.6)$$

This expression suggests that a two-sample analogue of the sign statistic might be $\sum_{i=1}^{m} \text{sgn}[X_i - \bar{z}(t)]$. However, we shall replace $\bar{z}(t)$ by $\hat{z}(t)$, the median of the $z_i(t)$. One obvious advantage of this alteration is that the null distribution of the resulting statistic

$$S(\mathbf{X}, \mathbf{Y}, t) = \sum_{i=1}^{m} \text{sgn}\,[X_i - \hat{z}(t)] \qquad (4.7)$$

is the same for all samples that have the same sample sizes m and n.

The null distribution of S
If N is even there are $N/2$ of the sgn $[z_i(t) - \hat{z}(t)]$ with value -1 and $N/2$ with value $+1$, and listing the exact permutation distribution of S is straightforward. In fact, $(S + m)/2$ has a hypergeometric distribution, and

$$\Pr(S = 2r - m) = \binom{N/2}{r}\binom{N/2}{m-r} \Big/ \binom{N}{m}$$

$r = 0, 1, 2, \ldots, m$, for $m < n$.

When N is odd, one of the sgn $[z_i(t) - \hat{z}(t)]$ has the value 0 which must be noted when listing the null distribution. The procedure remains straightforward.

Hypothesis testing
After establishing the null distribution of S, the hypothesis-testing procedure is simply to calculate observed $S(\mathbf{X}, \mathbf{Y}, \theta_0)$ when testing $H_0 : \theta = \theta_0$, and refer it to the null distribution. If the alternative hypothesis is $H_1 : \theta > \theta_0$, H_0 is rejected if observed S falls in the lower tail of the null distribution.

Example 4.3 Use the data of Example 4.2. Suppose we test H_0: $\theta_0 = 1.0$.
Then we have $H_1 : \theta > 1.0$

$$x_i : 2.71, 3.46, 4.13$$
$$y_j - \theta_0 : 3.85, 4.22, 4.64, 5.20$$

giving $\hat{z}(\theta_0) = 4.13$, the transformed values

$$x_i : -1, -1, 0$$
$$y_j - \theta_0 : -1, 1, 1, 1,$$

and the null distribution

$s:$	-3	-2	-1	0	1	2	3
$35\,\Pr(S = s):$	1	3	9	9	9	3	1

Observed $S = -2$ and $\Pr[S \leqslant -2] = 4/35$. At level $4/35$ this observed result is significant.

Confidence limits for θ

The null distribution of S is needed and we need to trace the values of $S(\mathbf{x}, \mathbf{y}, t)$ as a function of t for fixed $\mathbf{x}$ and $\mathbf{y}$. To cope with ties that occur as t varies, it is useful to spell out the definition of median $z_i(t)$ as follows:

$m + n$ odd : $\hat{z}(t)$ is the largest z'-value such that the number of $z_i(t)$ smaller than z' is $(m + n - 1)/2$.

$m + n$ even : $\hat{z}(t)$ is the mean of $\hat{z}_-(t)$ and $\hat{z}_+(t)$ where $\hat{z}_-(t)(\hat{z}_+(t))$ is the smallest (greatest) z' such that the number of $z_i(t)$ smaller than z' is $(m + n)/2$.

The setting of confidence limits is slightly different in the two cases, $m + n$ even or odd.

1. $m + n$ even ($m < n$). As t varies from $-\infty$ to $+\infty$, S varies between $-m$ and $+m$ in m steps of size 2. Suppose that $t = y_{((n-m)/2+1)} - x_{(m)} - \varepsilon$. Then the number of $z_i(t) < y_{((n-m)/2+1)}$ is $(n+m)/2$ and of these m are x-values. Thus $S = -m$. At $t = y_{((n-m)/2+1)} - x_{(m)} + \varepsilon$, the value of S is $-m + 2$. Proceeding in this way, we see that the t-values at which the jumps in S occur are $t_1 = y_{((n-m)/2+1)} - x_{(m)}, t_2 = y_{((n-m)/2+2)} - x_{(m-1)}, \ldots, t_m = y_{(n+m)/2} - x_{(1)}$.

Suppose we seek a $100\left(1 - 2r \bigg/ \binom{n+m}{m}\right)\%$ two-sided confidence interval. Then from the null distribution of S we find the value of S such that $\Pr(S \leqslant s) = r \bigg/ \binom{m+n}{n}$. Graphically, or by inspection of a tabulation, the t_j-value for the upper limit of the confidence interval is readily identified. The lower limit is found similarly.

2. $m + n$ odd ($m < n$). As t varies from $-\infty$ to $+\infty$, S varies from $-m$ to $+m$ in $2m$ steps of size 1. Following the type of enumeration illustrated for the case $m + n$ even, the t-values at which the jumps in S occur can be seen to be:

$$y_{(1)} - x_{(m)}, y_{(2)} - x_{(m)}$$
$$y_{(2)} - x_{(m-1)}, y_{(3)} - x_{(m-1)}$$

$$\vdots$$

$$y_{(n-1)} - x_{(1)}, y_{(n)} - x_{(1)}$$

An illustration ($m + n$ odd) follows.

Example 4.4 Use the data of Example 4.2. The 6 points of jump of S are:

$$
\begin{array}{ll}
(-3) & (0) \\
4.85 - 4.13 = 0.72 & 5.64 - 3.46 = 2.18 \\
(-2) & (1) \\
5.22 - 4.13 = 1.09 & 5.64 - 2.71 = 2.93 \\
(-1) & (2) \\
5.22 - 3.46 = 1.76 & 6.20 - 2.71 = 3.49 \\
(0) & (3)
\end{array}
$$

The values of S in the appropriate intervals are shown in brackets. From the null distribution of S shown in Example 4.3 we see that

$$\Pr\{-1 \leqslant S \leqslant +1\} = 27/35 = 0.7714$$

Therefore a 77% two-sided confidence interval for θ is:

$$(1.09, \quad 2.93)$$

Point estimation

From the descriptions above of the behaviour of S as a function of t for fixed $\mathbf{x}$ and $\mathbf{y}$, it is obvious that the point estimate of θ is

$$\text{median } (y) - \text{median } (x).$$

Large-sample calculations

In hypothesis testing we can use a normal approximation for the null distribution as explained in Section 4.3.2. In the case of the sign statistic the value of $\sigma^2(t)$ is

$$
\begin{cases}
1 & \text{for } m+n \text{ even, giving } \mathrm{var}(S(\mathbf{X}, \mathbf{Y}, \theta)) = mn/(m+n-1) \\
\dfrac{N-1}{N} & \text{for } m+n \text{ odd, giving } \mathrm{var}(S(\mathbf{X}, \mathbf{Y}, \theta)) = mn/(m+n)
\end{cases}
$$

For N large, the resulting test statistic is, for $m + n$ either even or odd, approximately equal to the usual test statistic arising in the analysis of a 2×2 contingency table.

Example 4.5 Suppose that the results of an X-sample of size 24 and a Y-sample of size 36 are used to test the null hypothesis that $\theta = 0$. If the common median is $\hat{z}(0)$ the results may be summarized in a 2×2 contingency table as follows:

	$< \hat{z}(0)$	$> \hat{z}(0)$	
X-sample	10	14	24
Y-sample	20	16	36
	30	30	

$$\mathrm{var}(S) = \frac{24 \times 36}{59}; \qquad \mathrm{s.d.}(S) = 3.827$$

Observed $S = -10 + 14 = 4$

$$\Pr\{S \geqslant 4 | H_0\} \simeq 1 - \Phi\left(\frac{3}{3.827}\right) = 1 - \Phi(0.784) = 0.216$$

Note the continuity correction of -1 in the numerator of the argument of Φ.

In the usual analysis of the 2×2 contingency table, using the normal approximation for the hypergeometric distribution arising in the 'exact' test we obtain, letting Q denote the entry in the top right-hand cell of the table,

$$E(Q) = 12$$
$$\mathrm{s.d.}(Q) = 1.913$$

giving the observed normal deviate, with the usual correction for continuity,

$$u = \frac{14 - 12 - 0.5}{1.913} = 0.784$$

identical to the result obtained above.

Consistency of the sign test

The common median $\hat{z}(t)$ may be regarded as an estimate of $\zeta_{0.5}(t)$, the median of the 'mixed' distribution function

$$[mF(z) + nF(z - \theta + t)]/N$$

As m and $n \to \infty$ with, say, $m/N \to \lambda$, the difference made to $\hat{z}(t)$ by omission of one of the X_i from its calculation is of order $1/N$. Hence, asymptotically we shall treat X_i and $\hat{z}(t)$ as independent and we obtain

$$E\{\mathrm{sgn}(X_i - \hat{z}(t))\} = 1 - 2F(\zeta_{0.5}(t)) + O(1/N) \qquad (4.8)$$

Arguing similarly that $X_i - \hat{z}(t)$ and $X_j - \hat{z}(t)$ are asymptotically independent, we conclude that

$$E\{S(\mathbf{X}, \mathbf{Y}, t)\}/m = \{1 - 2F(\zeta_{0.5} - (t))\} + O(1/N)$$
$$\mathrm{var}\{S(\mathbf{X}, \mathbf{Y}, t)\}/m = (\text{constant})/m + O(1/N)$$

from which the consistency of the sign test follows.

Asymptotic efficiency of the sign test

We consider the efficacy of the statistic S when $\theta = 0$. For this we need

$$\left[\frac{\partial E\{S(\mathbf{X}, \mathbf{Y}, t)\}}{\partial t}\right]_{t=0} \simeq -2m\{f(\zeta_{0.5}(0))\zeta'_{0.5}(0) + O(1/N)\} \qquad (4.9)$$

approximately for large m and n from (4.8), where

$$mF(\zeta_{0.5}(t)) + nF(\zeta_{0.5}(t) - t) = (m + n)/2 \qquad (4.10)$$

Differentiating both sides of the relation in (4.10), we obtain

$$mf(\zeta_{0.5}(t))\zeta'_{0.5}(t) + nf(\zeta_{0.5}(t) - t)(\zeta'_{0.5}(t) - 1) = 0$$

so that, if $f(\zeta_{0.5}(0)) \neq 0$, we have $\zeta'_{0.5}(0) = n/(m + n)$. Also, putting $t = 0$ in (4.10), we see that

$$F(\zeta_{0.5}(0)) = 1/2$$

that is $\zeta_{0.5}(0)$ is the median of the distribution F. Thus

$$\left[\frac{\partial E\{S(\mathbf{X}, \mathbf{Y}, t)\}}{\partial t} \right]_{t = 0} \simeq - \frac{2mn}{(m + n)} \{ f(\zeta_{0.5}(0)) + O(1/N^2) \} \qquad (4.11)$$

Since

$$\text{var } S(\mathbf{X}, \mathbf{Y}, 0) = mn/(N - 1)$$

the efficacy of the statistic S is

$$\frac{\left| \lim_{m,n \to \infty} (\partial E\{S(\mathbf{X}, \mathbf{Y}, t)\}/\partial t)_{t = 0} \right|}{\left(\dfrac{mn}{N} \right)^{1/2} \text{s.d.}(S(\mathbf{X}, \mathbf{Y}, 0))} = 2f(\zeta_{0.5}(0))$$

Example 4.6 Suppose that F has mean μ and variance σ^2. Then, in the case $\theta = 0$ we see from (4.2) that

$$\left(\frac{\partial E\{A(\mathbf{X}, \mathbf{Y}, t)\}}{\partial t} \right)_{t = 0} = \frac{mn}{N}$$

also s.d. $(A(\mathbf{X}, \mathbf{Y}, 0)) = \sqrt{(nm/N)}$, so that the efficacy of A is $1/\sigma$.

Thus the ARE of the tests based on S and on A is $4f^2(\zeta_{0.5}(0))\sigma^2$.

If F is a normal distribution, this ARE is $2/\pi$, coinciding with the value of the ARE of the S and A tests in the one-sample case.

Efficiency of estimation based on S

We have noted above that the point estimate of θ is the difference of the two medians, that is $\breve{\theta} = \text{median}(Y) - \text{median}(X)$. From results in Chapter 2, simple calculations show that asymptotically the distribution of $\breve{\theta}$ is normal with mean θ and

$$\text{var}(\breve{\theta}) = \frac{1}{4f^2(\zeta_{0.5}(0))} \left(\frac{1}{m} + \frac{1}{n} \right)$$

If, as in Example 4.6, $\text{var}(X) = \sigma^2$, the variance of the estimate $\bar{\theta} = \bar{Y} - \bar{X}$ is $\sigma^2 \left(\dfrac{1}{m} + \dfrac{1}{n} \right)$, giving the relative efficiency of estimation $1/(4\sigma^2 f^2(\zeta_{0.5}(0))$, equalling the ARE of testing.

4.3.4 The two-sample rank sum statistic

Referring to the expression (4.6) for the statistic $A(\mathbf{X}, \mathbf{Y}, t)$, an obvious modification to a rank statistic is to the statistic $\sum \text{Rank}\,(X_i - \bar{z}(t))$, where the $\text{Rank}\,(X_i - \bar{z}(t))$ is the rank in the combined sample of $z_i(t)$ values. However, since $\text{Rank}(X_i - \bar{z}(t)) = \text{Rank}(X_i)$ in the combined sample, we shall use

$$W(\mathbf{X}, \mathbf{Y}, t) = \sum_{i=1}^{m} \text{Rank}(X_i) - m(N+1)/2 \qquad (4.12)$$

the well-known Wilcoxon rank–sum statistic, except that the value $m(N+1)/2$ has been subtracted so that the null distribution of W has expectation 0.

The null distribution of W

Under H_0 the distribution of $\text{Rank}(X_i)$ is uniform on the integers $1, 2, \dots, N$, hence $E\{\text{Rank}(X_i)\} = (N+1)/2$.

The conditional null distribution of W, obtained according to the randomization procedure of Section 4.2, is in principle, easily enumerated. Since the null distribution of W is identical for all samples of the same size, inference using W is conditionally and unconditionally distribution-free. The null distribution of W has been tabulated for various values of m and n; see, for example, Lehmann (1975, Table B). A simple illustration follows.

Example 4.7 Suppose $m = 2$, $n = 3$, then the possible X-sample ranks are:

$$
\begin{array}{ll}
1,2 & W = -3 \\
1,3 & W = -2 \\
\;\;\vdots & \\
4,5 & W = +3
\end{array}
$$

and we obtain

w :	-3	-2	-1	0	1	2	3
$10\,\Pr(W=w)$:	1	1	2	2	2	1	1

In Example 4.7 the distribution of W is seen to be symmetric about 0. Such symmetry holds for all values of n because the distribution of W can be regarded for $N = 2k + 1$ as being generated by sampling at random without replacement from a population comprising elements whose values are

$$-k, -k + 1, \ldots, -1, 0, 1, 2, \ldots, k - 1, k$$

Since each possible sample occurs with the same probability, every sample of size m yielding

$$r_1 + r_2 + \ldots + r_m$$

has associated with it a sample yielding $-r_1 - r_2 - \ldots - r_m$. A similar argument applies when N is even.

The general formulae (4.4) can be applied to obtain the first two moments of W; in the present case we need the mean and variance of the uniform discrete distribution on the integers $1, 2, \ldots, N$, and they are respectively $(N + 1)/2$ and $(N^2 - 1)/12$. Hence we have

$$E(W) = 0$$
$$\text{var}(W) = mn(N + 1)/12$$

For large m and n, the distribution of W can be approximated by a normal distribution; the condition for the null distribution of A to be asymptotically normal is clearly satisfied for W. (See also Chapter 1 for information about asymptotic normality of rank statistics.) Readily available tables of W cover values of m and n between 3 and 10. For larger values of m and n, the normal approximation should be adequate for most practical purposes.

As an example, if $m = 8$, $n = 10$, $\Pr(W \leqslant -19) = 0.0506$. Using the normal approximation

$$\Pr(W \leqslant -19) \simeq \Phi\left\{ \frac{-19 + 0.5}{[(8 \times 10 \times 19)/12]^{1/2}} \right\} = \Phi(-1.644) = 0.0500$$

Hypothesis testing
Suppose that we test the hypothesis $H_0 : \theta = \theta_0$ against $H_1 : \theta = \theta_1 > \theta_0$. We shall consider the value of $E\{W(\mathbf{X}, \mathbf{Y}, \theta_0)|H_1\}$. Write

$$\text{Rank}(X_i) = 1 + \sum_{j \neq i = 1}^{m} V_{ji} + \sum_{r = 1}^{n} U_{ri}(t) \qquad (4.13)$$

where $V_{ji} = 1$ if $X_j < X_i$; 0 otherwise

$U_{ri}(t) = 1$ if $Y_r - t < X_i$; 0 otherwise

Then

$$E\{\text{Rank}(X_i)\} = \frac{m+1}{2} + n \Pr\{Y - t < X\}$$

$$= \frac{m+1}{2} + n \int F(x + t - \theta)f(x)\,\mathrm{d}x$$

in our model. Substituting in (4.12), we see that

$$E\{W(\mathbf{X}, \mathbf{Y}, \theta_0) | \theta = t\}$$

$$= \frac{m(m+1)}{2} + mn \int F(x + \theta_0 - t)f(x)\,\mathrm{d}x - m(N+1)/2 \quad (4.14)$$

which is clearly non-increasing with t since $F(x + \theta_0 - t)$ is non-increasing with t for fixed x. Thus, with $\theta_1 > \theta_0$,

$$E\{W(\mathbf{X}, \mathbf{Y}, \theta_0) | H_1\} < 0$$

and we reject H_0 if observed $W(\mathbf{X}, \mathbf{Y}, \theta_0)$ is 'sufficiently small'.

Example 4.8 Use the data of Example 4.2 to test $H_0 : \theta = 1.0$ against $H_1 : \theta > 1.0$. We have

$$x_i : 2.71\ (1), \quad 3.46\ (2), \quad 4.13\ (4)$$

$$y_j - \theta_0 : 3.85\ (3), \quad 4.22\ (5), \quad 4.64\ (6), \quad 5.20\ (7)$$

the numbers in brackets indicating the ranks. Thus

$$\text{observed } W(\mathbf{X}, \mathbf{Y}, \theta_0) = -5$$

From the null distribution of W, $\Pr(W \leqslant -5) = 2/35$, hence the observed value of W is significant at level $2/35$.

Confidence limits for θ

The effect of varying t in $W(\mathbf{x}, \mathbf{y}, t)$ is to shift the entire Y-sample relative to the X-sample. Consequently, the ranks of the X-sample values remain fixed as t varies except at points such that the order of an x_i, y_j pair is reversed, that is, at values $y_j - x_i$. Thus, as t varies from $-\infty$ to $+\infty$, $\sum_{i=1}^{m} \text{Rank}(X_i)$ varies between $m(m+1)/2$ and $N(N+1)/2 - n(n+1)/2$ in mn steps of size 1 at points $y_j - x_i$, $i = 1, 2, \ldots, m, j = 1, 2, \ldots, n$; and $W(\mathbf{x}, \mathbf{y}, t)$ varies between $-mn/2$ and $+mn/2$. Note that the values of W are either integers or integer multiples of $1/2$.

For a two-sided $100(1 - \alpha)\%$ confidence interval we find w_1 and w_2

such that $\Pr\{w_1 \leqslant W \leqslant w_2\} = 1 - \alpha$. If $d_{(1)}, d_{(2)}, \ldots, d_{(mn)}$ is the set of differences $y_j - x_i$ arranged in increasing order of magnitude, $r_1 = mn/2 - w_2, r_2 = mn - r_1 + 1$, then the confidence limits are $d_{(r_1)}, d_{(r_2)}$.

Example 4.9 Using the data of Example 4.2 the $y_j - x_i$ differences are as shown below with the values of W in brackets.

	(-6)		(0)
0.72		2.14	
	(-5)		(1)
1.09		2.18	
	(-4)		(2)
1.39		3.51	
	(-3)		(3)
1.51		2.74	
	(-2)		(4)
1.76		2.93	
	(-1)		(5)
2.07		2.49	
	(0)		(6)

From the null distribution of W, $\Pr\{-3 \leqslant W \leqslant 3\} = 1 - 8/35$, and by inspection of the table above the $100(1 - 8/35)\% = 77.1\%$ confidence interval for θ is $(1.39, 2.74)$.

Note that $r_1 = 6 - 4 = 2, r_2 = 12 - 2 + 1 = 11$.

Point estimation
In the light of the discussion above about the behaviour of $W(\mathbf{x}, \mathbf{y}, t)$ as t varies, the solution of the estimating equation

$$W(\mathbf{x}, \mathbf{y}, t) = 0$$

is clearly the median of the differences $y_j - x_i$. This estimating equation is used since $E\{W(\mathbf{X}, \mathbf{Y}, \theta)\} = 0$.
   In Example 4.9 the point estimate of θ is $(2.07 + 2.14)/2 = 2.105$.

Relation to the Mann–Whitney statistic
The *Mann–Whitney* two-sample statistic for testing $H_0: \theta = \theta_0$ is

$$U(\mathbf{X}, \mathbf{Y}, \theta_0) = \sum_{j=1}^{n} \sum_{i=1}^{m} U_{ji}(\theta_0)$$

where

$$U_{ji}(\theta_0) = \begin{cases} 1 & \text{if } Y_j - \theta_0 < X_i \\ 0 & \text{otherwise} \end{cases}$$

Referring to (4.13), we see after simple calculations that

$$W(\mathbf{X}, \mathbf{Y}, t) = U(\mathbf{X}, \mathbf{Y}, t) - \frac{mn}{2} \tag{4.15}$$

Note that $E\{U(\mathbf{X}, \mathbf{Y}, \theta)\} = mn \, \Pr(Y - \theta < X)$ and when $\theta = 0$, so that the X- and Y-distributions are identical under our present model, that $G(y) = F(y - \theta)$, then $E\{U(\mathbf{X}, \mathbf{Y}, 0)\} = mn/2$.

Since the Mann–Whitney statistic is, therefore, simply a linear function of the Wilcoxon statistic, the two statistics lead to identical inferences about θ.

Consistency of the W test
Rewriting (4.14), we have

$$E\{W(\mathbf{X}, \mathbf{Y}, t)\} = m(m+1)/2 + mn \int F(x + t - \theta)f(x) \, dx - m(N+1)/2$$

Hence for t close to θ, say $t = \theta + \Delta$,

$$E\{W(\mathbf{X}, \mathbf{Y}, t)\} \simeq m(m+1)/2$$

$$+ mn \int [F(x) + \Delta f(x)]f(x) \, dx - m(N+1)/2$$

$$= m(N+1)/2 + \Delta mn \int f^2(x) \, dx - m(N+1)/2$$

$$= \Delta mn \int f^2(x) \, dx \tag{4.16}$$

We need, also, $\text{var}\{W(\mathbf{X}, \mathbf{Y}, t)\}$, which we derive using the expression

$$W(\mathbf{X}, \mathbf{Y}, t) = \sum_{j=1}^{n} \sum_{i=1}^{m} U_{ji}(t) - mn/2$$

It is the sum of elements of the covariance matrix of $U_{11}(t), \ldots, U_{nm}(t)$. Many of these covariances are 0 because

$$\text{cov}\{U_{ji}(t), U_{rs}(t)\} = 0$$

unless $j = r$ or $i = s$. Let

$$q(t) = \Pr(Y - t < X) = \int F(x + t - \theta)f(x)\,dx.$$

Then

$$\text{cov}\,\{U_{ji}(t), U_{js}(t)\} = \Pr[Y_j - t < X_i, Y_j - t < X_s] - q^2(t)$$

$$= \int [1 - F(y - t)]^2 f(y - \theta)\,dy - q^2(t)$$

$$= w_1(t)$$

$$\text{cov}\,\{U_{ji}(t), U_{ri}(t)\} = \Pr[Y_j - t < X_i, Y_r - t < X_i] - q^2(t)$$

$$= \int [F(x + t - \theta)]^2 f(x)\,dx - q^2(t)$$

$$= w_2(t)$$

and

$$\text{var}\,(U_{ji}(t)) = q(t)(1 - q(t)) = \gamma(t)$$

Adding all the relevant variances and covariances of $U_{ij}(t)$ variables, we find

$$\text{var}\,\{W(\mathbf{X}, \mathbf{Y}, t)\} = mn[\gamma(t) + (m - 1)w_1(t) + (n - 1)w_2(t)] \quad (4.17)$$

As a check on the calculations of $E\{W(\mathbf{X}, \mathbf{Y}, t)\}$ and var $\{W(\mathbf{X}, \mathbf{Y}, t)\}$, we note that, for $t = \theta$,

$$E\{W(\mathbf{X}, \mathbf{Y}, \theta)\} = m(m + 1)/2 + mn \int F(x)f(x)\,dx - m(N + 1)/2 = 0,$$

which agrees with our previously derived result for the null distribution.

Further, $q(\theta) = 1/2$,

$$w_1(\theta) = \int [1 - F(y - t)]^2 f(y - \theta)\,dy - 1/4$$

$$= \int_0^1 (1 - u)^2\,du - 1/4 = 1/12,$$

and $w_2(\theta) = 1/12$ by a similar calculation. Therefore

$$\text{var}\,\{W(\mathbf{X}, \mathbf{Y}, \theta)\} = mn(m + n + 1)/12,$$

which agrees with the variance of the conditional and unconditional distribution of W.

Supposing that $m/N \simeq \lambda$, $n/N \simeq 1 - \lambda$ as $m, n \to \infty$, formulae (4.16) and (4.17) show that

$$E\{W(\mathbf{X}, \mathbf{Y}, t)\} \simeq C_1 N^2$$

$$\text{var}\{W(\mathbf{X}, \mathbf{Y}, t)\} \simeq C_2 N^3$$

from which the consistency of the W-test is readily established.

Efficiency and large-sample power of the W test
From the expression for $EW(\mathbf{X}, \mathbf{Y}, t)$ derived from (4.14) and used in deriving (4.16) we see that

$$\left(\frac{\partial E\{W(\mathbf{X}, \mathbf{Y}, t)\}}{\partial t} \right)_{t=\theta} = mn \int f^2(x)\, dx = mn\bar{f}$$

Hence the efficacy of W is

$$\lim_{\substack{m, n \to \infty \\ (m/N) \to \lambda}} \frac{mn\bar{f}}{\left(\dfrac{mn}{N} \right)^{1/2} [mn(m + n + 1)/12]^{1/2}} = \bar{f}\sqrt{12}$$

Thus, if F has variance σ^2 the ARE of the W and A tests is

$$12\bar{f}^2\sigma^2.$$

Example 4.10 If F is a normal distribution, with variance σ^2, it is readily established that $\bar{f} = 1/(2\sigma\sqrt{\pi})$, so that the ARE is $3/\pi$.

It is instructive also to calculate the power for large m and n when the alternative hypothesis is close to the null hypothesis. Suppose that we consider $H_1 : \theta_0 = \theta_0 + \delta/\sqrt{N}$. Then substituting in (4.16), putting $m \simeq \lambda N$, $n = (1 - \lambda)N$, we have

$$E\{W(\mathbf{X}, \mathbf{Y}, \theta_0)|H_1\} \simeq -\delta\lambda(1 - \lambda)N^{3/2}\bar{f}. \tag{4.18}$$

Similar substitutions in the expressions for $q(t)$, $w_1(t)$, and $w_2(t)$, and using the Taylor-type approximation appearing in the derivation of (4.16), enable one to obtain

$$\text{s.d.}\{W(\mathbf{X}, \mathbf{Y}, \theta_0)|H_1\} \simeq [\lambda(1 - \lambda)]^{1/2} N^{3/2} (1/\sqrt{12})(1 + O(1/\sqrt{n}))^{1/2}.$$

Using a normal approximation for the null distribution of W, a level-α test rejects H_0 in favour of H_1 if observed $W(\mathbf{X}, \mathbf{Y}, \theta_0)$ is greater than $u_\alpha[mn(N+1)/12]^{1/2}$. Assuming that the distribution of $W(\mathbf{X}, \mathbf{Y}, \theta_0)$ is approximately normal under H_1 as well as under H_0 (see Chapter 1), the power is approximately

$$1 - \phi\{u_\alpha - [\lambda(1-\lambda)]\sqrt{12\bar{f}}\}$$

Efficiency of estimation of θ

According to equation (1.5), Chapter 1, the large-sample variance of the point estimate of θ obtained as the solution, $\hat{\theta}$, of the estimating equation $W(\mathbf{x}, \mathbf{y}, t) = 0$ is

$$\text{var}(\hat{\theta}) \simeq \text{var}\, W(\mathbf{X}, \mathbf{Y}, \theta) \bigg/ \left[\frac{\partial E\{W(\mathbf{X}, \mathbf{Y}, t)\}}{\partial t} \right]_{t=\theta}^2 = \frac{N}{12mn\bar{f}^2}$$

Thus, relative to the estimator $\bar{Y} - \bar{X}$, the efficiency of $\hat{\theta}$ is $12\sigma^2\bar{f}^2$, if F has variance σ^2.

Large-sample calculations

When m and n are large, calculation of $W(\mathbf{X}, \mathbf{Y}, t)$ can be performed by first grouping the observations into classes to produce two frequency distributions. Then, taking all observations within a class interval to be uniformly distributed, an approximation to $W(\mathbf{x}, \mathbf{y}, t)$ can be calculated as explained in the following example.

Example 4.11 Suppose that the X- and Y-samples are grouped into frequency distributions as shown below:

X-sample		Y-sample	
Group	Frequency	Group	Frequency
5–10	2	14–18	1
10–15	4	18–22	4
15–20	9	22–26	8
20–25	7	26–30	11
25–30	6	30–34	12
30–35	2	34–38	6
		38–42	6
	$m = 30$	42–46	2
			$n = 50$

A cross-tabulation such as the one shown below is useful for listing the sums of $U_{ji}(t)$ values contributed by the observations in pairs of class intervals. The tabulation below is for $t = 0$; for other t-values the Y class limits are obtained by subtracting t from each of the limits shown above.

		\multicolumn{6}{c}{X}
		5–10	10–15	15–20	20–25	25–30	30–35
$Y-0$		2	4	9	7	6	2
14–18	1	0(0.8)	0.2(4)	7.0(9)	7(7)	6(6)	2(2)
18–22	4	0(0.8)	0(14.4)	3.6(36)	25.2(28)	24(24)	8(8)
22–26	8	0(0)	0(7.2)	0(70.2)	12.6(56)	46.8(48)	16(16
26–30	11	0(0)	0(0)	0(39.6)	0(77)	26.4(66)	22(22)
30–34	12	0(0)	0(0)	0(0)	0(50.4)	0(72)	14.4(2.4)
34–38	6	0(0)	0(0)	0(0)	0(1.0)	0(27.9)	0.3(12)
38–42	6	0(0)	0(0)	0(0)	0(0)	0(3.6)	0(10.8)
42–46	2	0	0	0	0	0(0)	0(0.9)

Where two intervals do not overlap their contribution to the sum of $U_{ji}(t)$ values is simply calculated; for example the $Y-0$ class 22–26 and the X-class 30–35 gives $8 \times 2 = 16$ $Y_i - 0$ values smaller than X_i, a contribution of 16 as entered in the table. The contribution from the $Y-0$ class 22–36 and X-class 15–20 is 0. Complication only arises when the intervals overlap. Take the $Y-0$ class 22–26 and X-class 20–25 with frequencies 8 and 7 respectively. We assume that the $8 \times 7 = 56$ pairs of values are uniformly distributed on the square in which $y-0$ varies from 22–26 and x varies from 20–25. A simple calculation gives the number of $Y_j - 0 - X_i$ differences less than 0 as $(4.5/20) \times 56 = 12.6$. The sum of cell entries in the cross-tabulation is 221.5, giving the observed $W(\mathbf{x}, \mathbf{y}, 0) = 221.5 - 750 = -528.5$.

To test the null hypothesis $\theta = 0$ we can use the value of W calculated above together with $\text{var}(W|H_0) = 10\,125 = (100.6)^2$. The observed result is, therefore, highly significant.

If we put $t = 10$, the $Y-10$ class intervals become $4-8, 8-12$, etc., and the corresponding contributions to $W(\mathbf{x}, \mathbf{y}, 10)$ are shown in brackets in the cross-tabulation above. These give $W(\mathbf{x}, \mathbf{y}, 10) = 744.6 - 750 = -5.4$.

Similar calculations for a selection of t-values gives

t	$W(\mathbf{x}, \mathbf{y}, t)$
0	-528.5
5	-296.6
7	-192.8
10	-5.4
12	96.5
15	288.2

Plotting the values of W against t, we obtain the points shown in Fig. 4.1; the straight line shown in this figure was drawn by eye. Note that the range of t values was selected to give W-values lying roughly between 0 ± 2 s.d.(W).

From the graph we can obtain:

(i) the point estimate of $\theta \simeq 10.2$; note that $\bar{y} - \bar{x} = 10.2$ calculated from the grouped data.

(ii) by finding the t-values corresponding to ± 1.645 s.d. $(W) = \pm 165.5$, approximately 90% two-sided confidence limits for θ are

$$(7.3, \quad 13.0)$$

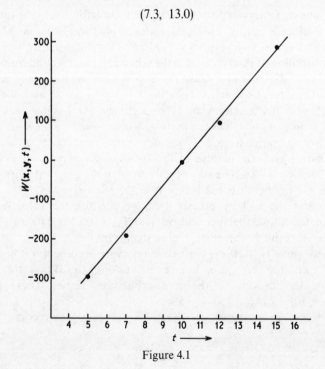

Figure 4.1

4.3.5 *Two-sample transformed rank statistics*

Let $R_i = \text{Rank}(X_i)$, where $\text{Rank}(X_i)$ is defined as in Section 4.3.4. Transform $R_i/(N+1)$ to $H_i = H[R_i/(N+1)]$, where H^{-1} is often taken to be a continuous distribution function. Then the statistic based on these transformed ranks, a modification of $W(\mathbf{X}, \mathbf{Y}, t)$ defined in (4.12), is

$$W_H(\mathbf{X}, \mathbf{Y}, t) = \sum_{i=1}^{m} H[R_i/(N+1)] - m\bar{H} \qquad (4.19)$$

where $\bar{H} = (1/N) \sum_{j=1}^{N} H(j/(N+1))$.

The null distribution of W_H

By the basic randomization procedure of Section 4.2, tabulation of the null distribution of W_H is, in principle, straightforward. Since the collection of numbers H_i, $i = 1, 2, \ldots, N$, is the same for every sample, the conditional and unconditional distributions of W_H coincide. Thus inference based on W_H is conditionally and unconditionally distribution-free.

Example 4.12 $m = 2$, $n = 3$, $H^{-1} = \Phi$, the standard normal distribution.

R_i:	1	2	3	4	5				
$H(R_i/6)$:	−0.967	−0.431	0	0.431	0.967				
w:	−1.398	−0.967	−0.536	0.431	0	0.431	0.536	0.967	1.39
$10\Pr[W_H = w]$:	1	1	1	1	3	1	1	1	1

In Example 4.12 the distribution of W_H is symmetric about 0 because the distribution function ϕ is symmetric with $\phi^{-1}(p) = -\phi^{-1}(1-p)$. In general, the distribution of W_H is not exactly symmetric about 0 but we have

$$E\{W_H(\mathbf{X}, \mathbf{Y}, \theta)\} = 0$$

$$\text{var}\{W_H(\mathbf{X}, \mathbf{Y}, \theta)\} = \frac{mn}{N(N-1)} \sum (H_i - \bar{H})^2$$

As $N \to \infty$, $(1/N) \sum_{j=1}^{N} H'(j/(N+1)) \to \int_0^1 H'(u)\,du$, so that for large N,

$$\text{var}\{W_H(\mathbf{X}, \mathbf{Y}, \theta)\} \simeq N\lambda(1-\lambda)\left[\int_0^1 H^2(u)\,du - \left(\int_0^1 H(u)\,du \right)^2 \right] \qquad (4.20)$$

Thus, if $H = \phi^{-1}$, one finds that $\text{var}\{W_H(\mathbf{X}, \mathbf{Y}, \theta)\} \simeq N\lambda(1-\lambda)$.

In this case $H = \phi^{-1}$, the resulting test statistic is called the van der

Waerden statistic; see, for example, Bradley (1968). Referring to Sections 2.3.3 and 2.3.4 it will be noted that the result obtained by the transformation Φ^{-1} is similar to the result that would be obtained if the observation X_i were replaced by $E(Y_{(r_i)})$, where $Y_{(r)}$ is the rth order statistic in a sample of size N from a standard normal distribution. The corresponding statistic is called the 'normal-scores' statistic; see also Example 2.14 and definition (2.44).

Hypothesis-testing and confidence limits

The hypothesis-testing procedure is straightforward, requiring calculation of the observed W_H and its referral, appropriately, to the null distribution of W_H.

For determination of exact confidence limits we could follow an argument like that used in Section 4.3 in connection with determining confidence limits based on the statistic $A(\mathbf{X}, \mathbf{Y}, t)$. However, since the conditional null distribution of W_H for testing any chosen value of t is invariant with respect to t, only the value of observed $W_H(\mathbf{x}, \mathbf{y}, t)$ varying with t, we can use the simpler procedure followed with $W(\mathbf{X}, \mathbf{Y}, t)$.

Example 4.13 Using the data of Example 4.2, $m = 3$, $n = 4$ and the inverse normal transformation, $H = \Phi^{-1}$, we have $H(1/8) = -1.1503$, $H(2/8) = 0.6745$, $H(3/8) = -0.3186$, etc., and straightforward enumeration gives

$$\Pr[-1.1503 \leqslant W_H \leqslant +1.1503] = 1 - 8/35$$

As t varies, observed $W_H(\mathbf{x}, \mathbf{y}, t)$ changes at $t = y_j - x_i$, $i = 1, \ldots, m$, $j = 1, 2, \ldots, n$. The following table shows the points at which W_H changes and the values of W_H in brackets.

	(−2.1434)		0
0.72		2.14	
	(−1.8248)		(0.4758)
1.09		2.18	
	(−1.5062)		(0.7944)
1.39)		2.51	
	(−1.1503)		(1.1503)
1.51		2.74	
	(−0.7944)		(1.5062)
1.76		2.93	
	(−0.4758)		(1.8248)
2.07		3.49	
	0		(2.1434)

From this table, a $100(1 - 8/35)\% = 77.1\%$ confidence interval for θ is $(1.39, 2.74)$.

This result coincides with the result of Example 4.9. However, in general W and W_H will not give identical results; this can be seen by noting that while the functions of t, W and W_H have 'jumps' at the same t values the jumps are of equal size in W but of unequal size in W_H.

The efficacy of W_H

We need an expression for $E\{W_H(\mathbf{X}, \mathbf{Y}, t)\}$ in order to find the efficacy of W_H. For this purpose we note that

$$\{\text{Rank}(X_i)|X_i = x\} = 1 + \#(X_j < x; j \neq i) + \#(Y_k - t < x), \quad (4.21)$$

from which it readily follows that

$$E\{\text{Rank}(X_i)|x\} = 1 + (m - 1)F(x) + nF(x + t - \theta)$$

$$\text{var}\{\text{Rank}(X_i)|x\} = (m - 1)F(x)[1 - F(x)]$$
$$+ nF(x + t - \theta)[1 - F(x + t - \theta)]$$

For m and n large, $m \simeq \lambda N$, $n \simeq (1 - \lambda)N$, we then have

$$E\left\{\frac{\text{Rank}(X_i)}{(N \div 1)}\bigg| X_i = x\right\} \simeq \lambda F(x) + (1 - \lambda)F(x + t - \theta)$$

$$\text{var}\left\{\frac{\text{Rank}(X_i)}{(N + 1)}\bigg| X_i = x\right\} \leqslant \frac{1}{4N}$$

from which

$$E\left[H\left\{\frac{\text{Rank}(X_i)}{N + 1}\right\}\bigg| X_i = x\right] \simeq H\{\lambda F(x) + (1 - \lambda)F(x + t - \theta)\}$$

and

$$E\left\{H\left(\frac{\text{Rank}(X_i)}{N + 1}\right)\right\} \simeq \int H\{\lambda F(x) + (1 - \lambda)F(x + t - \theta)\}f(x)\mathrm{d}x$$

$$\partial E\left\{H\left(\frac{\text{Rank}(X_i)}{N + 1}\right)\right\}\bigg/ \partial t \simeq \int H'\{\lambda F(x) + (1 - \lambda)F(x + t - \theta)\} \times$$
$$\times (1 - \lambda)f(x + t - \theta)f(x)\mathrm{d}x$$

giving

$$\left(\frac{\partial E\{W_H(\mathbf{X}, \mathbf{Y}, t)\}}{\partial t}\right)_{t = \theta} \simeq \frac{mn}{N}\int H'\{F(x)\}f^2(x)\mathrm{d}x \quad (4.22)$$

Combining (4.22) and (4.20), the efficacy of W_H is

$$e(W_H) = \left| \int H'\{F(x)\} f^2(x)\,dx \right| \bigg/ \left\{ \int_0^1 H^2(u)\,du - \left[\int_0^1 H(u)\,du \right]^2 \right\}^{1/2}$$

(4.23)

As a check, putting $H(u) = u$, $H'(u) = 1$ in (4.23) reproduces the result (4.18).

4.3.6 'Robust' transformations in the two-sample case

Suppose that the values $Z_i(t) = X_i$, $i = 1, 2, \ldots, m$, $z_{m+j}(t) = Y_j - t$, $j = 1, 2, \ldots, n$, are transformed to $\psi(Z_k(t))$, $k = 1, 2, \ldots, N$, and define the statistic $A_\psi(\mathbf{X}, \mathbf{Y}, t)$ by

$$A_\psi(\mathbf{X}, \mathbf{Y}, t) = \sum_{i=1}^m \psi\left(\frac{X_i - c}{d} \right) - \frac{m}{N} \sum_{k=1}^N \psi\left(\frac{Z_k(t) - c}{d} \right) \quad (4.24)$$

Typically $\psi(u)$ is to be chosen to be continuous and monotonic in u, and such that its effect is to dilute the influence of outlying values of the $Z_k(t)$.

According to the basic randomization argument of Section 4.2 conditionally distribution-free inference about θ is possible with the statistic $A_\psi(\mathbf{X}, \mathbf{Y}, t)$. The arguments are essentially the same as those for the statistic $A(\mathbf{X}, \mathbf{Y}, t)$ elaborated in Section 4.3.2. Possible forms of ψ are those mentioned in a similar context in Section 2.3.6.

The location adjustment, c, and the scale factor, d, appearing in (4.24) can depend on t without the conditional inference argument being affected. As we have noted before, in Chapter 2, the commonly suggested ψ transformations are sensitive to the choice of scale factor and a 'best' choice is far from clear. Possible choices are

$$c = \text{median } (Z_i(t))$$

$$d = \text{median } (|Z_i(t) - c|)$$

Hypothesis testing
Application of the suggested transformations in hypothesis testing is straightforward, as shown in the following example.

Example 4.14 (See also Example 4.2.)

$$x : 2.71, 3.46, 4.13$$

$$y : 4.85, 5.22, 5.64, 6.20$$

We consider testing $H_0 : \theta = 1$ against $H_1 : \theta > 1$.

$$c = \text{median}\,(2.71, \ldots, 5.20) = 4.13$$

$$d = \text{median}\,(|2.71 - 4.13|, \ldots, |5.20 - 4.13|) = 0.51$$

$$\psi\left(\frac{x_i - c}{d}\right) : -0.884, \; -0.576, \; 0$$

$$\psi\left(\frac{y_j - 1 - c}{d}\right) : -0.268, \; 0.088, \; 0.462, \; 0.781$$

Observed $A_\psi = -1.460 - (-0.170) = -1.29$

Under H_0, $\Pr(A_\psi \leqslant -1.29) = 2/35$, by simple enumeration, hence H_0 is rejected at the 5.7% level.

Point estimation and confidence limits

The point estimate of θ based on A_ψ is the solution of the estimating equation

$$A_\psi(\mathbf{X}, \mathbf{Y}, t) = 0 \tag{4.25}$$

In principle, determination of exact confidence limits can be achieved by the type of argument used to find exact confidence limits based on A, as explained in Section 4.3.2. However, the computations are much more tedious and complications can arise through the possible non-monotic behaviour of A_ψ as a function of t. In large samples this is not a serious problem and two-sided confidence limits for θ can be obtained, using a normal approximation for the distribution of A_ψ, by solving

$$A_\psi(\mathbf{x}, \mathbf{y}, t) = \pm u_\alpha \sigma(t)(N/mn)^{1/2}, \tag{4.26}$$

where $\psi_i = \psi((Z_i(t) - c)/d)$ and $\sigma^2(t) = (1/N)\sum_{i=1}^{N}\psi_i^2 - (\sum\psi_i/N)^2$. The argument leading to (4.26) is exactly the same as that used to give (4.5).

4.3.7 *Multiplicative models*

Suppose that the positive random variables X and Y are such that the distribution of Y is the same as the distribution of ρX, interest being in estimating the multiplying factor ρ. Since the distributions of $\ln Y$ and $\ln \rho + \ln X$ are identical, the problem of inference about ρ can be regarded as similar to inference about a location-shift parameter.

Another class of alternatives that can also be reduced to the location shift type is

$$Y \text{ distributed like } X^\alpha$$

where, again Y and X are positive random variables. Here we have

$$\ln(\ln Y) \text{ distributed like } \ln\alpha + \ln(\ln X)$$

One of the practical advantages of using distribution-free techniques based on signs or ranks in connection with data generated by distributions of the type in question, arises through the tendency of experiments to round off observations, in particular to record 0 for very small positive values. If the proportion of such 0 values is not large, they do not cause any difficulty in analyses by rank or sign transformations.

4.4 Proportional hazards (Lehmann alternative)

In life testing and the analysis of survival data, a model that arises naturally and is frequently used is

$$1 - F(x) = (1 - G(x))^\alpha \tag{4.27}$$

For the X-population the hazard rate at x, also called the *instantaneous death rate*, is

$$\lim_{\Delta \to 0} \frac{\text{Probability of 'death' in } (x, x + \Delta x)}{\text{Probability that } X > x} = \frac{f(x)}{1 - F(x)} = h_X(x)$$

For the Y population we have, similarly, the hazard rate $h_Y(y) = g(y)/[1 - G(y)]$. From (4.27) we see that

$$f(x) = \alpha(1 - G(x))^{\alpha - 1} g(x)$$

so that the X-population hazard rate is

$$h_X(x) = \frac{f(x)}{1 - F(x)} = \frac{\alpha(1 - G(x))^{\alpha - 1} g(x)}{(1 - G(x))^\alpha} = \frac{\alpha g(x)}{(1 - G(x))} = \alpha h_Y(x) \tag{4.28}$$

The proportionality of the hazard rates in the X and Y populations exhibited by (4.28) explains the alternative terminology for the model (4.27).

4.4.1 *The Wilcoxon statistic and inference about α*

In the present context it will be convenient to use the Wilcoxon or the Mann–Whitney statistic in the more standard form, rather than the form $W(\mathbf{x}, \mathbf{y}, t)$ as in (4.12) and (4.15). Let

$$U_{ji} = \begin{cases} 1 & Y_j < X_i \\ 0 & \text{otherwise} \end{cases}$$

and put

$$U(\mathbf{X}, \mathbf{Y}) = \sum_{j=1}^{n} \sum_{i=1}^{m} U_{ji} \qquad (4.29)$$

This $U(\mathbf{X}, \mathbf{Y})$ is the Mann–Whitney statistic, and if $W(\mathbf{X}, \mathbf{Y})$ denotes the sum of ranks of the X-sample after pooling the two samples, we have

$$W(\mathbf{X}, \mathbf{Y}) = m(m + 1)/2 + U(\mathbf{X}, \mathbf{Y})$$

The expectation of $U(\mathbf{X}, \mathbf{Y})$ is readily seen to be

$$E\{U(\mathbf{X}, \mathbf{Y})\} = mn\mathrm{Pr}(Y < X)$$

$$= mn \int G(y)f(x)\,\mathrm{d}x = mn \int (1 - F(y))g(y)\,\mathrm{d}y$$

Clearly $E\{U(\mathbf{X}, \mathbf{Y})\} = (1/2)mn$ if $G = F$.

In the model (4.27) we have

$$\mathrm{Pr}(Y < X) = \int [1 - G(y)]^{\alpha} g(y)\,\mathrm{d}y = 1/(1 + \alpha) \qquad (4.30)$$

again, putting $\alpha = 1$, so that $F = G$, gives $\mathrm{Pr}(Y < X) = 1/2$.

Since we have a simple expression for $E\{U(\mathbf{X}, \mathbf{Y})\}$ in terms of the parameter α, it appears that U is a natural statistic to use for inference about α. Up to a point this is true, but, as we shall explain, there are difficulties.

Testing $H_0 : \alpha = 1$ against $H_1 : \alpha < 1$
Under H_0, $E(U) = mn/2$ and under H_1, $E(U) > mn/2$; hence we reject H_0 if the observed U is sufficiently large. As the distributions F and G are identical under H_0, the actual test procedure can use the null distribution of W as it is described in Section 4.3.4. The observed U (or W) can be referred to tabulated percentage points of the null distribution of the U (or W) statistic, or a suitable normal approximation to this distribution can be used.

Testing $H_0 : \alpha = \alpha_0 \neq 1$ against $H_1 : \alpha < \alpha_0$.
In the application of the Wilcoxon test to the case of a specified location-shift alternative, θ_0, it is possible to transform the Y-sample to an adjusted sample, $Y_1 - \theta_0, Y_2 - \theta_0, \ldots, Y_n - \theta_0$, which can be

regarded as being generated by the same distribution as the X-sample. Hence the exact distribution of $W(\mathbf{Y}, \mathbf{X}, \theta_0)$ could be derived by the permutation argument of Section 4.2. In the proportional-hazards model such a simple transformation producing, by the permutation argument, an exact distribution of W under H_0 when $\alpha_0 \neq 1$ does not seem possible. In principle it is possible to express all the joint probabilities of the U_{j_i} that are needed for the distribution of U in terms of α_0, but the task is prohibitively tedious.

If we take m and n to be sufficiently large for the distribution of U to be approximately normal, we need only the first and second moments of U expressed in terms of α_0.

For general α we can follow steps like those leading to the expression (4.17):

$$q = \Pr(Y < x) = 1/(\alpha + 1); \quad \gamma = \alpha/(\alpha + 1)^2$$

$$w_1 = \text{cov}(U_{ji}, U_{hs}) = \Pr[Y_j < X_i, Y_i < X_j] - q^2$$

$$= \int [1 - F(y)]^2 g(y) \, dy - q^2$$

$$= 1/(2\alpha + 1) - 1/(\alpha + 1)^2$$

$$= \alpha^2/[(2\alpha + 1)(\alpha + 1)^2]$$

$$w_2 = \text{cov}(U_{ji}, U_{ri}) = 1 - 2\alpha/(\alpha + 1)$$

$$+ \alpha/(\alpha + 2) - 1(\alpha + 1)^2$$

$$= \alpha/[(\alpha + 2)(\alpha + 1)^2]$$

giving

$$\text{var}\{W(X, Y)\} = mn\alpha/(\alpha + 1)^2 [1 + (n - 1)\alpha/(2\alpha + 1)$$

$$+ (m - 1)/(\alpha + 2)] \tag{4.31}$$

For a specified α_0 in H_0, we can then obtain $E(W|H_0)$ and $\text{var}(W|H_0)$ simply by substituting α_0 for α in the expressions (4.30) and (4.31).

Example 4.15

$$X : \quad 0.315 \quad 1.062 \quad 1.357 \quad 0.004$$

$$Y : \quad 0.062 \quad 0.574 \quad 0.578 \quad 0.654 \quad 1.226 \quad 0.087$$

Testing $H_0 : \alpha = 0.6$ against $\alpha < 0.6$.

$$E(U|H_0) = 15, \text{ from (4.30)}$$
$$\text{var}(W|H_0) = 19.78 = (4.448)^2, \text{ from (4.31)}$$
$$\text{Observed } U = 13.$$

Since observed U differs from $E(U|H_0)$ by less than one s.d. $(U|H_0)$. we accept H_0.

Confidence limits for α

Using a normal approximation for the distribution of U, with (4.30) and (4.31), an approximate $100(2\beta - 1)\%$ confidence interval $(a_1 a_2)$ for α can be found by solving the following equation in α for a_1 and a_2.

$$\text{Observed } U - mn/(\alpha + 1) = \pm u_\beta \text{ s.d. } (U) \tag{4.32}$$

where u_β is an appropriate normal deviate, and $[\text{s.d}(U)]^2$ is given by the r.h.s. of (4.31).

Example 4.16 Using the data in Example 4.15, graphical solutions of the equations (4.32) give, with $u_\beta = 1.645$, approximate 90% two-sided confidence limits for α as $(0.27, 2.9)$. The point estimate of α is $a = mn/(\text{observed } U) - 1 = 0.85$.

Efficiency of U for α

The point estimate $\tilde{\alpha} = (mn/U) - 1$ has large-sample variance approximately $\{(\alpha + 1)^4/(m^2 n^2)\} \text{var}(U)$. For comparison, consider the maximum-likelihood estimates $\hat{\alpha}$ of α in the special case $1 - F(x) = e^{-\alpha x/\mu}$, $1 - G(y) = 1 - e^{-y/\mu}$. We then have $\hat{\alpha} = \bar{Y}/\bar{X}$ with large-sample var $\tilde{\alpha} \simeq \alpha^2(N/mn)$. Thus at $\alpha = 1$, $\text{var}(\hat{\alpha})/\text{var}(\tilde{\alpha}) \simeq 3/4$.

4.4.2 *The 'log-rank test' and inference about* α

The test which we are about to describe is important on several counts:

(i) It is a valid test of equality of F and G in general, not only in the proportional hazards case.

(ii) In the proportional-hazards case it has certain optimality properties.

(iii) The test can be generalized to deal with cases where some observations are censored; in certain types of applications, notably studies of survival, such a test is very useful.

We begin with a description of the test of the case $\alpha = 1$ and no censoring of observations. Since the terminology is useful, and this test is often applied in survivorship analysis, we shall think of the observed variables as life times. Then the pooled sample values $z_1, z_2, \ldots, z_N$ are the times at which deaths occur, in either the X-sample or the Y-sample. Let $C_X(z)$ be 1 if the death at time z is an X-

death, 0 otherwise. Let m_j, n_j denote the numbers in the X- and Y-samples respectively that are still alive at time $z_i - 0$.

Bearing in mind that we are considering the proportional-hazards model, it seems natural to develop a test procedure based directly on calculations of conditional probabilities of events at the observed death times. More specifically, if the hazard rates in the two populations are identical, then the conditional probability $\Pr[C_X(S_j) = 1 | m_j, n_j]$ is simply equal to the proportion of the X-sample 'at risk' at time $z_j - 0$. Thus

$$\Pr[C_X(z_j) = 1 | m_j, n_j] = \frac{m_j}{m_j + n_j}$$

another way of expressing this result is

$$e_j = E(C_X(z_j) | m_j, n_j) = m_j / (m_j + n_j)$$

The proposed test statistic is

$$U = \sum C_X(z_j) - \sum e_j = m - \sum e_j \qquad (4.33)$$

As usual, when the null hypothesis implies that F and G are identical it is possible to evaluate the exact randomization distribution of U by the argument of Section 4.2, as illustrated in Example 4.17

Example 4.17

X :	0.7,	1.5,	4.7	$(m = 3)$
Y :	0.9,	2.3		

z_j	m_j	n_j	$c_X(z_j)$	e_j
0.7	3	2	1	$3/5 = 0.6$
0.9	2	2	0	$2/4 = 0.5$
1.5	2	1	1	$2/3 = 0.6$
2.3	1	1	0	$1/2 = 0.5$
4.7	1	0	1	$1/1 = 1.0$
	Totals		3	3.26

Inspection of the column headed e_j will show that, whatever assignment of three of the five observations is made to the X-sample, the form of $\sum e_j$ is always

$$\sum e_j = U_5/5 + U_4/4 + \ldots + U_1/1$$

and the full set of $\binom{5}{3}$ realizations of $(U_5, U_4, \ldots, U_1)$ is as follows:

U_5	U_4	U_3	U_2	U_1	$\sum e_j$	$\sum q_j$	$u/(\sum q_j)^{1/2}$
3	2	1	0	0	1.433	0.712 22	1.856
3	2	1	1	0	1.933	0.962 22	1.087
3	2	2	1	0	2.266	0.962 22	0.748
3	3	2	1	0	2.516	0.899 72	0.510
3	2	1	1	1	2.933	0.962 22	0.068
3	2	2	1	1	3.266	0.962 22	− 0.272
3	3	2	1	1	3.516	0.899 72	− 0.545
3	2	2	2	1	3.766	0.712 22	− 0.908
3	3	2	2	1	4.016	0.649 72	− 1.261
3	3	3	2	1	4.350	0.427 50	− 2.065

A simple calculation shows that $E(\sum e_j) = 3 = m$; hence $E(U) = 0$. The observed U for our particular example is $3-3.266$; reference to the list of all possible $\sum e_j$ shows that the result is not significant.

Further inspection of Example 4.17 should convince the reader that it is a fairly simple matter to describe the exact distribution of any U_j for general m and n. In fact

$$\Pr(U_j = s) = \binom{j}{s}\binom{N-j}{m-s} \bigg/ \binom{N}{M}, \quad s = 0, 1, 2, \ldots, M \quad (4.34)$$

indicating that U_j has a hypergeometric distribution. Using (4.34) and noting that $U = m - \sum_{j=1}^{N} U_j/j$, we have

$$E(U) = m - \sum_{j=1}^{N} M_j/N_j = 0$$

as verified in the example above.

Another way of checking $E(U) = 0$ is by noting that

$$E\left\{\sum C_X(z_j) | m_1, n_1, \ldots, m_N, n_N\right\} = E\left\{\sum C_X(z_j) | \mathbf{m}, \mathbf{n}\right\} = \sum e_j$$

whence

$$E\left\{\sum C_X(z_j)\right\} = E\left(\sum e_j\right)$$

but $\quad \sum C_X(z_j) = m$, a constant, giving $E\left(\sum e_j\right) = M$.

A useful result concerning var (U) can be obtained by similar reasoning. By repeated application of the basic relation

$$\text{var}(Y) = E\{\text{var}(Y|X)\} + \text{var}\{E(Y|X)\} \qquad (4.35)$$

$$\text{var}\left\{\sum C_X(z_j)|\mathbf{m}, \mathbf{n}\right\} = \sum_{j=1}^{N} m_j n_j/(m_j + n_j)^2 \qquad (4.36)$$

Putting $q_j = m_j n_j/(m_j + n_j)^2$, $j = 1, 2, \ldots, N$, we have

$$\text{var}\left\{\sum C_X(z_j) - \sum e_j|\mathbf{m}, \mathbf{n}\right\} = \sum q_j.$$

Also

$$E\left\{\sum C_X(z_i) - \sum e_j|\mathbf{m}, \mathbf{n}\right\} = 0$$

and since $\sum C_x(z_i) = \text{constant} = m$,

$$\begin{aligned}
\text{var}(U) &= \text{var}\left(\sum e_j\right) \\
&= E(\textstyle\sum q_j) + \text{var}\{E(U|m, n)\} \\
&= E(\textstyle\sum q_j) \qquad (4.37)
\end{aligned}$$

When m and n are large one may treat U as being approximately normally distributed with variance given by the observed value of $\sum q_j$. Some justification for such a procedure is provided by (4.37). in that the observed $\sum q_j$ may be regarded as an estimate of var (U). Alternatively one may argue conditionally on $\mathbf{m}, \mathbf{n}$, in which case (4.37) is not needed. Approximate normality of U cannot be established by elementary means.

The effect of using such a normal approximation is illustrated by the last two columns in the tabulation appearing with Example 4.17. The only two values of $|U/(\sum q_j)^{1/2}|$ that exceed the 10% point (1.282) of a standard normal distribution are those in the first and last rows of the table.

4.4.3 Conditional likelihood and the log-rank test

In a paper on regression models and life tables Cox (1972) uses a conditional likelihood approach to develop a class of test procedures of which the two-sample log-rank test is a special case. The argument is similar to that used in Section 4.4.3, for, conditionally on m_i, n_i, individuals being 'exposed' at $z_i - 0$ the probability that the 'death' at z_i is that the individual as observed is

$$\frac{\alpha^{C_X(z_i)}}{\alpha m_i + n_i}$$

Each failure contributes a similar factor to the overall conditional likelihood whose logarithm is

$$L(\alpha) = \log \alpha \sum_{i=1}^{N} C_X(z_i) = \sum_{i=1}^{n} \log(\alpha m_i + n_i)$$

with

$$\frac{\partial \log L(\alpha)}{\partial \alpha} = \log \sum_{i=1}^{N} \left\{ \frac{C_X(z_i)}{\alpha} - \frac{m_i}{(\alpha m_i + n_i)} \right\} = U(\alpha) \qquad (4.38)$$

The relation (4.38) suggests the use

$$U = \sum \left\{ C_X(z_i) - \frac{m_i}{m_i + n_i} \right\}$$

as a statistic for testing that $\alpha = 1$, for, if $\alpha = 1$, then, the solution $\hat{\alpha}$ of $U(\alpha) = 0$ should not 'differ significantly' from 1.

4.4.4 *The log-rank test and censored observations*

For the null hypothesis $F = G$, modification of the log-rank test for the case where some observations are censored is straightforward. By censoring we understand that some observations may be known to exceed certain values, but their exact values are unknown. Typically, 'right-censored' observations of this sort occur in survival studies where, for example, a patient who is still alive is lost from a study by the action of some agency such as moving to another town, or dying accidentally.

The modification to the log-rank test is simply to calculate the conditional expectations e_j, $\sum e_j$, and the relevant conditional variances, using numbers m_i', n_1', these being interpreted exactly like the m_i, n_i except that they are numbers exposed after subtraction of numbers of deaths and of numbers lost.

In practical applications, observations like survival times are often rounded off to the nearest day or month or year. This means that for any z_i the values of c_i and $c_x(z_i)$ may be greater than one. No new principle is needed to develop a test for the conditionally expected number of X-deaths at z_i is just

$$e_i' = \left(\frac{m_i'}{m_i' + n_i'} \right) c_i$$

and the conditional variance is

$$q_i' = n_i' m_i' - \frac{c_i(n_i' + m_i' - c_i)}{(n_i' + m_i' - 1)(n_i' + m_i')^2}.$$

The test statistic is

$$U = \sum_{i=1}^{k} \left\{ C_X(z_i) - \left(\frac{m_i'}{m_i' + m_i'} \right) c_i \right\}$$

where k is the number of distinct death times.

In principle it is possible to obtain a randomization null distribution for U, in the manner of Example 4.17, but there are problems associated with deciding exactly how the random allocations should be made. In particular one may ask whether randomization should be performed disregarding censoring or whether the randomization should be performed subject to the numbers of censored observations in the X-sample also being held fixed.

However, the difficulty may be ignored when m and n are moderately large, in which case the calculated conditional mean and variance of U can be used, assuming approximate normality of the distribution of U.

Example 4.18 We use a collection of data on remission times of leukemia, published by Freireich *et al.* (1963) and used by Gehan (1965) to illustrate a method for dealing with censored observations, and also by Cox (1972); the * indicates a censored result.
X-sample (exposed to drug 6-MP): 6*, 6, 6, 6, 7, 9*, 10*, 10, 11*, 13, 16, 17*, 19*, 20*, 22, 23, 25*, 32*, 32*, 34*, 35*.
Y-sample (control): 1, 1, 2, 2, 3, 4, 4, 5, 5, 8, 8, 8, 8, 11, 11, 12, 12, 15, 17, 22, 23.

Failure time (z_i)	m_i'	n_i'	c_i'	$C_X(z_i)$	e_i'	q_i'
1	21	21	2	0	1.0000	0.4878
2	21	19	2	0	1.0500	0.4860
3	21	17	1	0	0.5526	0.2472
4	21	16	2	0	1.1352	0.4772
5	21	14	2	0	1.2000	0.4659
6	21	12	3	3	1.9092	0.6508
7	17	12	1	1	0.5862	0.2426
8	16	12	4	0	2.2856	0.8707
10	15	8	1	1	0.6522	0.2268
11	13	8	2	0	1.2381	0.4481
12	12	6	2	0	1.3333	0.4183
13	12	4	1	1	0.7500	0.1875
15	11	4	1	0	0.7333	0.1956
16	11	3	1	1	0.7857	0.1684
17	10	3	1	0	0.7692	0.1775
22	7	2	2	1	1.5556	0.3025
23	6	1	2	1	1.7143	0.2041
Totals				9	19.2509	6.2570

From the table above $U = 10.25$ with variance $= 6.26$.

4.5 Dispersion alternatives

Elementary discriptions of distributions are often concerned mainly with measures of location and of dispersion. Thus in considering alternatives other than these already discussed in Sections 4.3 and 4.4 it is natural to think of dispersion alternatives. Unfortunately, it seems possible to develop simple exact tests only if distributions are assumed to be equally located. Otherwise the two samples have to be 'aligned' by the subtraction of an estimate of the difference in location from one set of sample values. A test of equality of dispersion based on such aligned sample is generally not exact.

Another non-trivial difficulty is associated with the question of interpreting the result of a test of equality of dispersion if it is significant. In a general way every test of dispersion may be thought of as a test of equality of a certain measure of dispersion. However, it is not always obvious just what that measure is; in particular, it may be difficult to devise and use a point estimate of the relevant population value.

With these difficulties in mind we shall mention just two possible approaches to the testing of equality of dispersion.

4.5.1 *A randomized exact test of dispersion*

Rearrange the X-sample values in random order to give $x'_1, x'_2, \ldots, x'_m$ and do likewise for the Y-sample to give $y'_1, y'_2, \ldots, y'_n$. Now let

$$W_1 = |x'_1 - x'_2|, \quad W_2 = |x'_3 - x'_4|, \ldots, \quad W_k = |x'_{2k-1} - x'_{2k}|$$

$$V_1 = |y'_1 - y'_2|, \quad V_2 = |y'_3 - y'_4|, \ldots, \quad V_l = |x'_{2l-1} - x'_{2l}|$$

where $k(l)$ is the largest integer such that $2k \leqslant m$ $(2l \leqslant n)$.

One interpretation of the hypothesis of equal dispersion is that the W- and V-samples are from identically located populations. Now any one of the exact tests of location developed in earlier sections can be applied to the W- and V-samples. A word about terminology: this test is called randomized because of the preliminary random ordering of the X- and Y-samples (two statisticians will not necessarily obtain the same answer for the same test procedure). This randomization is not to be confused with the randomization of Section 4.2.

The problem of interpretation is emphasized by remarking that the X-dispersion parameter implicitly defined by a test procedure such as that outlined above is a location parameter of the distribution of $X_1 - X_2$, where X_1 and X_2 are identically distributed like X. This location parameter could be median of $|X_1 - X_2|$ or $E|X_1 - X_2|$.

A great advantage of the randomized test described above is that equality of location of the X- and Y-distributions need not be assumed. An obvious disadvantage is that it is randomized; randomized procedures are not popular with practitioners of statistics.

4.5.2 *Comparing interquartile ranges*

The interquartile range is perhaps one of the conceptually simplest measures of dispersion. Clearly, it bears a close analogy to the median in the problem of location. For a continuous population the interquartile range always exists and is unique.

If we wish to test the null hypothesis that two populations are identical against the alternative that the populations have the same median but different interquartile ranges, a simple test is possible. Let $z_{0.25N}$ and $z_{0.75N}$ denote the sample lower and upper quartiles in the pooled X- and Y-samples. Then the randomization distribution of the statistic $Q_X = \#$ (X observations between $z_{0.25N}$ and $z_{0.75N}$) is easily evaluated; it is a hypergeometric distribution. For large N values, $E(Q_X|H_0) = 0.5m$

$$\text{var}(Q_X|H_0) = 0.5mn/(N-1),$$

hence the significance of an observed value of Q_X is readily assessed; see Westenberg (1948).

When the assumption of equal location of the two populations cannot be made, separate point estimates of the interquartile ranges of the two populations can be found and compared in terms of their estimated variances; for a discussion of estimation of the variance of the sample interquartile range see Section 3.3.

4.5.3 Rank test for dispersion

Several rank or rank-like tests of dispersion exist for equally located distributions and are described in various texts, for example, Gibbons (1971). To illustrate a typical approach consider the Mood (1954) test statistic

$$M = \sum_{i=1}^{m} [\text{Rank}(X_i) - (N+1)/2]^2$$

where $\text{Rank}(X_i)$ is the rank of X_i in the pooled X- and Y-samples.

It is clearly easy in principle to enumerate the exact null distribution of M under randomization; in this distribution,

$$E(M) = (m(n^2 - 1)/12$$
$$\text{var}(M) = mn(N+1)(N^2 - 4)/180$$

EXERCISES

4.1 The following X- and Y-samples were drawn at random from populations differing only in location, by an amount Δ.

$$X: \quad 6.474, \quad 4.573, \quad 5.799$$
$$Y: \quad 6.560, \quad 7.699, \quad 9.084, \quad 8.539$$

Using the mean statistic and the basic randomization procedure obtain an exact $100(1 - 4/35)\%$ confidence interval for Δ.

4.2 Tabulate the exact null distribution of the two sample sign statistic for sample sizes $m = 3$, $n = 4$ as in Exercise 4.1.

Using the data of Exercise 4.1 and the two sample sign statistic test $H_0 : \Delta = 3.0$ against $H_1 : \Delta > 3.0$ at the level $\alpha = 4/35$.

4.3 In the two-sample problem let $m = 1, n = 2$. Then the sign statistic for testing equality of medians is

$$S(0) = \mathrm{sgn}(X_1 - \hat{Z})$$

where $\hat{Z}$ is the median of X_1, Y_1, Y_2.

Obtain the exact distribution of $S(0)$ in terms of the X and Y distributions, F and G.

4.4 Consider the sign statistic defined as in Exercise 4.3 for general m and n such that $N = m + n$ is odd. Then

$$\mathrm{sgn}(X_i - \hat{Z}) = -1, 0, +1,$$

if

$$\# (X_j > X_i, j \neq i) + \# (Y_r > X_i) >, =, < (N - 1)/2.$$

Conditioning on $X_i = x$, express $P\{\mathrm{sgn}(X_i - \hat{Z}) = -1 | X_i = x\}$ in terms of F and G.

Hence, or otherwise express $E\{S(0)\}$ in terms of F and G.

4.5 Suppose that $m = 2$ observations are drawn at random from a $R\{0, 1\}$ population and $n = 2$ from an $R\{0.1, 1.1\}$ population.

Tabulate the exact distribution of the Wilcoxon rank sum statistic $W(\mathbf{X}, \mathbf{Y}, 0) = \sum \mathrm{Rank}(X_i) - 5$ where $\sum \mathrm{Rank}(X_i)$ denotes the sum of the ranks of the observations from the $R\{0, 1\}$ population.

Note: $R\{a, b\}$ denotes a continuous uniform distribution with density $f(x) = 1/(b - a)$ for $a \leqslant x \leqslant b$, and 0 elsewhere.

4.6 Check the first two moments of the distribution obtained in Exercise 4.5 against values given by the formulae in Section 4.4, Chapter 4.

4.7 In an experiment to examine the effect of a certain diet on growth rate of laboratory animals the following weights of randomly selected control and special diet fed animals were observed after a set time.

Control : 396, 409, 371, 367, 392 (X observations)
Diet fed : 434, 405, 440, 441, 399 (Y observations)

Assume that the effect of diet is multiplicative, that is, the distribution of Y is the same as the distribution of γX where γ is the multiplicative effect. Obtain a point estimate and a $100\beta\%$ confidence interval for γ based on the Wilcoxon rank sum statistic, with β as close as practicable to 0.95.

4.8 Assuming the multiplicative model of Exercise 4.7, and letting $g(y)$ be the probability density function of the Y distribution, derive the following large sample approximate formula for the point estimate $\tilde{\gamma}$, of γ, based on the Wilcoxon rank sum statistic:

$$\text{var}(\tilde{\gamma}) \simeq \frac{\gamma^2}{12\bar{g}_*^2}(1/m + 1/n)$$

where

$$\bar{g}_*^2 = \int y\,g^2(y)\,\mathrm{d}y$$

4.9 The following two frequency distributions summarize X and Y samples from populations that differ only in location.

Class lower limit	100	105	110	115	120	125	130
X-frequency	2	15	37	21	3		
Y-frequency	1	8	39	70	38	6	

Plot an approximate graph of the Wilcoxon rank sum statistic $W(\mathbf{X}, \mathbf{Y}, t)$ against t assigning the same (mid-rank) value to all observations within the same interval, and plotting at values of t that are integral multiples of the class width. Hence obtain a point estimate of the location shift parameter θ, and an approximate 90% confidence interval for θ.

CHAPTER 5

Straight-line regression

5.1 The model and some preliminaries

We shall consider independent continuous random variables $Y_1, Y_2, \ldots, Y_n$ that are observed at values $x_1, x_2, \ldots, x_n$ of a non-random variable X. If θ_j is a location parameter of $Y_j, j = 1, 2, \ldots n$, the straight-line regression problem is specified by

$$\theta_j = \alpha + \beta x_j, \qquad j = 1, 2, \ldots, n$$

where α and β are parameters. For the most part our objective is to make inferences about α and β from observations on $Y_1, Y_2, \ldots, Y_n$.

In the rather general setting outlined above, we may take the distribution function of Y_j to be $F_j(y), j = 1, 2, \ldots, n$, and, if no specialization of these distribution functions is introduced, it may be argued that the only sensible location parameter to consider is the median, that is, θ_j is taken to be the median of Y_j. As for restrictions on the F_j, they will be influenced by the view one takes of the straight-line regression problem. One view is that it is a matter of comparing locations of distributions indexed by $x_1, x_2, \ldots, x_n$. Such a view takes the problem to be an extension of the two-sample problem, and essentially displays no interest in the parameter α. All the interest centres in β, and in these circumstances a natural restriction is that the distributions F are identical except for location. Associated with this type of restriction there is a body of techniques for inference about β only. Another view is of simultaneous estimation of the locations of $Y_1, \ldots, Y_n$ through estimation of α and β. In such a formulation it is essential to specify just what location parameter is being estimated for each Y_j. This places us in much the same position as location estimation in the one-sample case. Recalling our previous discussions it will be apparent that we shall have to assume every Y_j to be symmetrically distributed about $\alpha + \beta x_j$.

5.2 Inference about β only

A number of techniques associated with the notions of 'testing for trend' or 'testing for randomness' become available for inference about β only if we assume that the distributions F_j are identical except for location. Basically the idea is that, under this assumption, the differences

$$D_j(\beta) = Y_j - \beta x_j, \qquad j = 1, 2, \ldots, n$$

are identically and independently distributed. Thus a test of trend of the $D_j(\beta)$ is a test that β is the true parameter value.

Extending the permutation argument used in the two-sample case, we see that the basic method of constructing a distribution-free test is to consider any permutation of the observed $D_j(\beta)$ values as possible and occurring with the same probability. Under such a scheme it is possible to tabulate the exact conditional null distribution of any statistic defined as a function of the $D_j(\beta)$ values.

Example 5.1 Consider testing $H_0 : \beta = 1$ using the data below.

$x_j : -2$	-1	0	1	2
$y_j : -1.35$	-1.55	-48	1.55	2.21
$D_j(1) : 0.65$	-0.55	-0.48	0.55	0.21

Take as the test statistic $T^*(\beta) = \sum \operatorname{sgn}(x_j)\, D_j(\beta)$. There are 120 permutations of the $D_j(1)$ values and the 5 permutations giving the largest values of $T^*(1)$ are

					$T^*(1)$
-0.55	-0.48	0.21	0.55	0.65	2.23
-0.55	-0.48	0.55	0.21	0.65	1.89
-0.55	-0.48	0.65	0.55	0.21	1.79
-0.55	0.21	-0.48	0.55	0.65	1.54
-0.48	0.21	-0.55	0.55	0.65	1.47

The values of $T^*(1)$ are shown in the last column above. The five smallest $T^*(1)$ values are obtained by reversing the order of each permutation.

Now, the observed $T^*(1) = 0.66$, a result that is not significant at the $100(10/120)\%$ level.

When two or more of the x_i values coincide care must be taken to list all possible permutations. The following example should be noted.

Example 5.2. Test $\beta = 1.2$ using the following data:

$$x_j : -2 \qquad 1 \qquad 1 \qquad 2$$
$$y_j : -1.35 \qquad 1.55 \qquad 0.98 \qquad 2.21$$
$$y_j - 1.2\,x_j : \quad 1.05 \qquad 0.35 \qquad -0.22 \qquad -0.19$$

Consider the statistic $T(\beta) = \sum x_i D_i(\beta)$

Permutation				$T(1)$
1.05	0.35	-0.22	-0.19	-2.35
1.05	0.35	-0.19	-0.22	-2.38
1.05	-0.22	0.35	-0.19	-2.35
1.05	-0.22	-0.19	0.35	-1.81
1.05	-0.19	0.35	-0.22	-2.38
1.05	-0.19	-0.22	0.35	-1.81
0.35	1.25	-0.22	-0.19	-0.25

etc.

There are not $4! = 24$ distinct $T(1)$ values, only 12, in this particular example, each of them occurring twice. They are.

$$-2.38 \qquad -0.25 \qquad 1.91$$
$$-2.35 \qquad 0.99 \qquad 2.00$$
$$-1.81 \qquad 1.34 \qquad 2.61$$
$$-0.28 \qquad 1.46 \qquad 2.70$$

A general class of statistics may be defined by

$$T(\beta, \psi, H; \mathbf{Y}, \mathbf{x}) = \sum_{j=1}^{n} \psi(x_j) H[D_j(\beta)] \qquad (5.1)$$

For notational convenience some of the arguments in $T(\beta, \psi, H; \mathbf{Y}, \mathbf{x})$ will occasionally be suppressed when there is no risk of confusion.

The transformed 'residuals' $H[D_j(\beta)]$ play a role rather like the deviations $(X_j - \theta)$ in the one-sample case or $Y_j - \theta$ in the two-sample case. Considerations as to the form of H apply here as they did in the one- and two-sample problems. An extra ingredient here is $\psi(x_j)$, with $\psi(x_j) = \mathrm{sgn}(x_j)$ being used in Example 5.1. It seems clear that a sensible choice of $\psi(x_j)$ should be such that a trend in the $D_j(\beta)$ values should produce a numerically large value of T. We shall see that, given H, it is usually possible to find a 'best' ψ. It should be noted that, such a 'best' ψ produces, for a given H, a best T only within the class of statistics defined by (5.1).

As is often the case, the best-known classical parametric procedures provide useful pointers to sensible choices of ψ and H, and also indicate obvious restrictions that may have to be placed on the x_j values. Recall, briefly, the case where the Y_j have expectations $\alpha + \beta x_j, j = 1, 2, \ldots, n$, and identical finite variance σ^2, and suppose, as we shall assume throughout this chapter, unless stated otherwise, that $\sum x_j = 0$. Well-known theory shows that

$$T(\beta) = \sum_{j=1}^{n} x_j D_j(\beta)$$

is 'best' for this case. Further, if β is estimated by the value of b satisfying the estimating equation

$$T(\beta) = \sum x_j (y_j - bx_j) = 0$$

then the variance of the estimate of β is $\sigma^2 / \sum x_j^2$.

This result shows that for consistency of estimation of β by $b = \sum x_j y_j / \sum x_j$ we must have $\sum x_j^2 \to \infty$ as $n \to \infty$. Thus we must exclude cases when, for example, all additional $|x_j|$ values tend to 0 as n is increased. This is also obviously dictated by elementary considerations that apply to the general statistic in (5.1) As we shall see, the best choice $\psi(x_j) = x_j$ in classical least squares, is also a best choice for other models and choices of H.

5.2.1 *Inference based on* $T(\beta) = \sum x_j D_j(\beta)$

The statistic

$$T(\beta, \mathbf{x}) = \sum x_j D_j(\beta) \tag{5.2}$$

can be regarded as the straight-line regression analogue of the mean statistic A of Chapter 2. It can be regarded as the primitive statistic, suggested by the method of least squares from which the statistic in (5.1) is derived by suitable generalization. It is obviously important in its own right since it is the best in the case where $E(Y_j) = \alpha + \beta x_j$ and $\text{var}(Y_j) = \sigma^2 < \infty$.

Hypothesis testing

To test the hypothesis $H_0 : \beta = \beta_0$, we evaluate the null distribution, of $T(\beta_0, \mathbf{x})$ conditional on $y_j - \beta_0 x_j, j = 1, 2, \ldots n$, fixed. This null distribution is obtained by listing all $n!$ permutations of the values $y_j - \beta_0 x_j, j = 1, 2, \ldots, n$, and calculating the value of T for each of them. Each such value has the same probability, $1/n!$, in the null distribution. The test of H_0 is then effected by referring the observed value of T to the null distribution, appropriate account being taken of whether the alternative hypothesis is one- or two-sided.

The work of enumerating the exact conditional null distribution of T increases rapidly with n, and one often has to resort to using the first two moments of the distribution in carrying out the test of hypothesis. Write $d_j = y_j - \beta_0 x_j, j = 1, 2, \ldots, n$, and put

$$\bar{d} = \sum d_j/n$$
$$v = \sum (d_j - \bar{d})^2/n$$

Let D_j be the random variable generated in position x_j by the permutation. Then, from elementary theory, as in Section 1.8.3,

$$E(D_j) = \bar{d}, \text{var}(D_j) = v, \text{cov}(D_j, D_k) = -v/(n-1), j \neq k.$$

Using these formulae we have

$$E(T) = \sum x_j \bar{d}$$

and

$$\text{var}(T) = v\sum x_j^2 - \sum_{j \neq k} x_j x_k v/(n-1)$$

$$= v\sum x_j^2 - \frac{v}{(n-1)}\{(\sum x_j)^2 - \sum x_j^2\}$$

Since we take $\sum x_j = 0$, we obtain, more simply,

$$E(T) = \bar{d}\sum x_j = 0$$
$$\text{var}(T) = (nv/(n-1))\sum x_j^2 \qquad (5.3)$$

Approximation of the distribution of T by a normal distribution will be possible if $\text{var}(Y_j)$ is finite and, as $n \to \infty$,

$$\max x_j^2/\sum x_j^2 \to 0$$

An illustration of the use of such an approximation is given in Example 5.3.

Confidence limits for β

Consider a two-sided $100(1 - 2r/n!)\%$ confidence interval for β; r will be chosen such that $2r/n!$ is close to one of the conventionally used small probabilities 0.10, 0.05, etc. The confidence interval will be constructed by finding the set of possible β values for which the appropriate null hypothesis is accepted.

For a selected possible value b of β, the hypothesis test is in principle carried out by listing all $n!$ possible values of the statistic $T(b)$ and noting that a typical member of this collection is

$$T_q(b) = \sum_{i=1}^{n} x_i(y_{n_i} - x_{n_i}b)$$

The subscript n_i for y and x in this expression indicates that we are dealing with a permutation of the $(y_i - bx_i)$ values; the letter q refers to the particular permutation, and it is also useful to write $T_q(b)$ as

$$T_q(b) = \sum_q x_i(y_{n_i} - bx_{n_i})$$

One of these $T_q(b)$ is the actually observed value $T(b)$ and we denote the set of all permutations excluding the observed one by Q'.

Let $N(b)$ denote the number of $T_q(b), q \in Q'$, values that are smaller than the observed $T(b)$. If $N(b) > r$ and $N(b) < n! - r$, then the hypothesis $\beta = b$ is accepted and b belongs to the confidence interval for β. In order to set the confidence limits we have, therefore, to examine the value of $N(b)$ as b varies from $-\infty$ to $+\infty$. Represent $N(b)$ as

$$N(b) = \sum_{q \in Q'} I[\sum x_i(y_i - bx_i) - \sum_q x_i(y_{n_i} - bx_{n_i}) > 0]$$

$$= \sum_{q \in Q'} I\left[\frac{\sum_{i=1}^{n} x_i(y_i - y_{n_i})}{\sum_{i=1}^{n} x_i(x_i - x_{n_i})} > b \right]$$

assuming that $\sum_{i=1}^{n} x_i(x_i - x_{n_i}) > 0$

Thus the confidence limits are the rth smallest and rth largest of all slope estimates

$$b_q = \frac{\sum_{i=1}^{n} x_i(y_i - y_{n_i})}{\sum_{i=1}^{n} x_i(x_i - x_{n_i})} \tag{5.4}$$

if $N(b)$ changes by 1 at each of the b_q values. This will be true if all x_i values are distinct. Equality of two or more x-values is possible and its effect is clearly that we do not have $n! - 1$ distinct values of b_q. In turn, this means that the value of $N(b)$ jumps by more than 1 at some of the b_q values. The implication of these facts is as follows: if there are $n! - 1$

distinct values of b_q, the confidence coefficient can be selected from the numbers $(1 - 1/n!)$, $(1 - 2n!)$, ... If there are not $n! - 1$ distinct values, the possible exact confidence coefficients are a subset of the numbers $(1 - 1/n!)$, $(1 - 2/n!)$, ... With n large it will still be possible, in general, to set an exact confidence coefficient close to one of the conventional values.

Example 5.3

$$x_i : \quad -3 \qquad -1 \qquad 1 \qquad 3$$
$$y_i : \quad -1.65 \qquad -1.25 \qquad 0.34 \qquad 4.17$$

The $4! - 1 = 23$ values of b_q calculated according to (5.4) are as follows, in order of magnitude:

0.2000	0.7950	0.9700	1.1900
0.3983	0.8453	1.0050	1.2558
0.4975	0.8600	1.0361	1.3550
0.5967	0.9000	1.0575	1.5417
0.6849	0.9262	1.1050	1.9150
0.7000	0.9525	1.1683	

Taking $r = 1, 2$, we find the following confidence intervals for β:

$$r = 1 : 100(1 - 2/24)\% = 91.7\% : (0.2000, 1.9150)$$
$$r = 2 : 100(1 - 4/24)\% = 83.3\% : (0.3983, 1.5417)$$

The least-squares point estimate is $\hat{\beta} = 0.9525$, which it will be noted, is the median of the b_q values listed above. If the x_i are symmetrically distributed about 0, the median of the b_q values always coincides with $\hat{\beta}$.

Applying the 'usual' normal theory in which the Y_i are taken to be normally distributed with the same variance, the 83% confidence limits for β are obtained by invoking the t-distribution with 2 degrees of freedom; the results are

91.7%:	0.9525 ± 0.8623
83.3%:	0.9525 ± 0.6112

Listing values of b_q can be prohibitively tedious for moderately large values of n, and the normal approximation of the conditional distribution of T can be used to simplify the calculation; this process must not be confused with using the standard normal-theory model!

Using the formulae (5.3), an application of Fieller's theorem gives the two sided $100(1 - 2\gamma)\%$ confidence limits (two-sided) for β as the values of b satisfying

$$(n-1)^{1/2}\left[\sum x_i y_i - b\sum x_i^2\right]$$
$$= \pm u_\gamma \left\{\sum(y_i - \bar{y})^2 - 2b\sum x_i y_i + b^2\sum x_i^2\right\}^{1/2}\left(\sum x_i^2\right)^{1/2} \qquad (5.5)$$

where u_γ is an appropriate normal quantile.

Example 5.4 We illustrate the use of (5.5) with the data of Example 5.3, although n is rather small.

$$\sum x_i^2 = 20, \sum x_i y_i = 19.05, \sum(y_i - \bar{y})^2 = 21.141\,475$$

and for an 83% confidence interval $u_\gamma = 1.38$.

The solutions of (5.5) are 0.44 and 1.46, giving an interval that agrees quite well with the exact result.

Denoting the usual least squares-estimate of β by $\hat{\beta} = \sum x_i y_i / \sum x_i^2$, we can write

$$\sum x_i(y_i - x_i b) = (\hat{\beta} - b)\sum x_i^2$$
$$\sum(y_i - \bar{y} - bx_i)^2 = \sum(y_i - \bar{y} - \hat{\beta}x_i)^2 + (\hat{\beta} - b)^2\sum x_i^2$$
$$= (n-2)s^2 + (\hat{\beta} - b)^2\sum x_i^2$$

where s^2 is the usual 'residual mean square'. Then the solution of (5.5) can be seen to be the solution of

$$(\hat{\beta} - b)^2\sum x_i^2\left[1 - u_\gamma^2/(n-1)\right] = s^2 u_\gamma^2(n-2)/(n-1)$$

giving

$$b = \hat{\beta} \pm u_\gamma\left(\frac{n-2}{n-1}\right)^{1/2}\left[1 - u_\gamma^2/(n-1)\right]^{-1/2}\left\{s/(\sum x_i^2)^{1/2}\right\} \qquad (5.6)$$

If the distribution of residuals were known to be normal with variance σ^2, the confidence limits would be given by (5.6) with the factor $K = u_\gamma[(n-2)/(n-1)]^{1/2}[1 - u_\gamma^2/(n-1)]^{1/2}$ replaced by $t_{n-2}(\gamma)$. The following table is instructive.

	$\sigma = 0.05$		$\gamma = 0.025$	
n	K_γ	$t_{n-2}(\gamma)$	K_γ	$t_{n-2}(\gamma)$
5	2.505	2.353	8.530	3.182
10	1.854	1.860	2.440	2.306
20	1.729	1.734	2.135	2.101
∞	1.645	1.645	1.960	1.960

For $n \geqslant 10$ the agreement between K_γ and t_{n-2} is very good; it breaks down for smaller n and as γ decreases.

Consistency and efficiency of T

The unconditional expectation of $T(b)$ is

$$E(T(b)) = \sum x_i[E(Y_i) - bx_i]$$

It is defined only if every $E(Y_i)$ is finite and, in that case we can take $E(Y_i) = \alpha + \beta x_i$. Therefore,

$$E(T(b)) = (\beta - b)\sum x_i^2$$

The variance of $T(b)$ is finite only if $\text{var}(Y_i) = \sigma^2 < \infty$, in which case $\text{var}(T(b)) = \sigma^2 \sum x_i^2$. Thus a test based on T will be consistent if $1/\sum x_i^2 \to 0$ as $n \to \infty$.

If the distribution of Y_j is $N(\alpha + \beta x_j, \sigma^2)$, the point estimate $\hat{\beta}$ obtained by solving $T(b) = 0$ is the maximum-likelihood estimate of β. The variance of $\hat{\beta}$ is $\sigma^2/\sum x_i^2$, hence the efficiency of T, relative to various statistics based on transformed residuals, will be low for heavy-tailed residual distributions.

5.2.2 *Transformations of the $D_j(\beta)$ to ranks*

Inference about β only can be regarded as inference about location differences so that procedures for β can be thought of as extensions of two-sample procedures. Pursuing this line of thought we can consider as an alternative to T the statistic

$$T_H(\beta, \mathbf{x}) = \sum x_j H(D_j(\beta))$$

where H is some suitable transformation of the $D_j(\beta)$ values. In particular

$$T_R(\beta, \mathbf{x}) = \sum x_i \text{Rank}(D_j(\beta))$$

can be regarded as an analogue of the Wilcoxon two-sample rank-sum statistic.

Hypothesis testing

We need the conditional null distribution of $T_R(\beta, \mathbf{x})$ for a specified β derived by the basic permutation procedure described earlier. Since the $D_j(\beta)$ values are replaced by their ranks, the null distribution is not only somewhat easier to enumerate but is also invariant with respect to β. Thus, for a given set of x-values the null distribution can be tabulated once and for all. Practically, this is not of great benefit unless we restrict attention to equally spaced x_j values, in which case the statistic T_R is a linear function of the Spearman rank correlation coefficient. Since the null distribution of T_R is invariant with respect to F, the test using T_R is also unconditionally distribution-free.

Example 5.5 (This example illustrates the method of listing the null distribution of T_R.) We use the data of Example 5.3 and consider the hypothesis $H_0 : \beta = 1.2$.

x_i:	-3	-1	1	3
y_i:	-1.65	-1.25	0.34	4.17
$y_i - 1.2x_i$:	1.95	-0.05	-0.86	0.57
Rank($y_i - 1.2x_i$):	4	2	1	3

Observed $T_R = -4$.

The null distribution of T_R is easily enumerated by listing the 24 permutations of the numbers 1, 2, 3, 4 and calculating the corresponding value of T_R, thus

x_i	-3	-1	1	3	T_R
	1	2	3	4	10
	1	2	4	3	8

etc.

giving the following distribution of T_R.

t:	-10	-8	-6	-4	-2	0	2	4	6	8	10
$24\,\mathrm{Pr}(T_R = t)$:	1	3	1	4	2	2	2	4	1	3	1

Thus the observed T_R is not at all extreme; in fact, $\mathrm{Pr}\{|T_R| \geqslant 4\} = 18/24$.

Listing the distribution of T_R is, of course, prohibitively tedious for $n \geqslant 6$, but in hypothesis testing one need only list the extreme values. Thus, with $n = 5$ for a one-sided test at the approximately 5% level one only has to list 5 or 6 permutations. For larger values of n it may be possible to use a normal approximation for the distribution of T_R.

Normal approximation of the null distribution of T_R
The argument giving formulae (5.3) for $E(T)$ and var(T) also gives $E(T_R)$ and var(T_R) with the d_j values replaced by the ranks $1, 2, \ldots, n$. Thus v in (5.3) becomes $(n^2 - 1)/12$ and we have

$$E(T_R) = 0$$
$$\mathrm{var}(T_R) = n(n + 1)\sum x_i^2/12 \qquad (5.7)$$

According to Section 1.8.4 the distribution of T_R, suitably norma-lized, is asymptotically normal as $n \to \infty$ if $\max x_i^2 / \sum x_i^2 \to 0$. It will be noted (Exercise 5.5) that the distribution of T_R is symmetric about 0 if the x_i are symmetrically positioned about 0. Hence the normal approximation may be quite good for moderately small n in the symmetric case, depending of course, on the actual spacing of the x-values.

Example 5.6 Let $n = 5$ with x_i values $-2, -1, 0, 1, 2$, giving $\sum x_i^2 = 10$, s.d.$(T_R) = 5$. Enumeration of the 5 permutations of 1, 2, 3, 4, 5 giving the largest values if T_R shows that $\Pr(T_R \geqslant 9) = 5/120 = 0.0416$. The normal approximation gives

$$\Pr(T_R \geqslant 9) \simeq 1 - \Phi(8.5/5) = 0.0446$$

Point estimation
Since

$$E\{T_R(\beta, \mathbf{x})\} = 0$$

an application of the method of moments suggests taking as an estimate of β the solution of the estimating equation

$$T_R(b, \mathbf{x}) = \sum x_j \operatorname{Rank}(y_j - bx_j) = 0 \qquad (5.8)$$

As b increases the residuals $y_j - bx_j$ for $x_j > 0$ decrease, hence their ranks do not increase; a similar remark applies to residuals for $x_j < 0$. Consequently, $T_R(b, x)$ is a non-increasing (step) function of b.

As b varies from $-\infty$ to $+\infty$ the value of T_R only changes when the ranking of residuals changes. Unless three or more observations are collinear, an event with zero probability, these changes occur whenever b coincides with one of the pairwise slopes $(y_j - y_i)/(x_j - x_i)$. There are $n(n - 1)/2$ of these if all x_i values are distinct. If n is even and the x's are symmetrically placed about 0 with the values $-n + 1, -n + 3, \ldots -1, 1, \ldots, n - 1$, the maximum value of T_R is $n(n^2 - 1)/6$. Thus it is clear that, in general, the steps in the graph of T_R against b are not of equal height. Therefore the solution of (5.8) is not necessarily the median of the pairwise slopes. In practice, a graph of T_R against b suitably smoothed near $T_R = 0$ is helpful in deciding on a point estimate of β. For most practical purposes smo-othing by eye should suffice. The justification for smoothing is that the step function is actually an estimate of a smooth function.

Example 5.7 The data of Examples 5.3 and 5.4 give the following

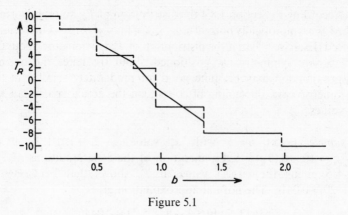

Figure 5.1

pairwise slopes, the values of T_R in the respective b intervals being shown in parentheses:

(10) 0.2 (8) 0.4975 (4) 0.795 (2) 0.97 (−4) 1.355 (−8) 1.915 (−10)

Figure 5.1 shows a graph of T_R against b, and according to the smoothing indicated on the graph we take the point estimate of β to be 0.94.

Confidence limits

Since the null distribution of T_R is invariant with respect to β, the determination of confidence limits is simpler than it is with the use of T in Section 5.2.1. Suppose that $\Pr(t_1 \leqslant T_R \leqslant t_2) = \gamma$. Then, to determine a $100\,\gamma\%$ confidence interval for β we need only scan the list of 'pairwise slopes', arranged in order of magnitude and the lower (upper) limit is the smallest (largest) value of b such that observed $T_R \leqslant t_2 (\geqslant t_1)$. These values will be just to the right (left) of one of the pairwise slopes.

Example 5.8 Refer to Example 5.5 and the list of pairwise slopes. From the null distribution of T_R given in Example 5.3 we have $\Pr(-8 \leqslant T_R \leqslant 8) = 22/24 = 0.917$. Therefore 91.7% confidence interval for β is (0.2, 1.915).

This result coincides with the 91.7% confidence interval given in Example 5.3; generally this will not happen. Note also, that we cannot obtain an 83.3% confidence interval in this case, because $\Pr(T_R = 8) = \Pr(T_R = -8) = 3/24$.

With larger values of n the values of t_1 and t_2 can usually be found with the normal approximation discussed above. When n is so large that evaluation of all $\binom{n}{2}$ pairwise slopes is impractical, an approximation to the graph of T_R against b, obtained by evaluating T_R at a judiciously selected set of values of b, should suffice for graphical determination of the confidence limits. These are given by the intersections of the graph T_R and the horizontal lines with abscissae t_1 and t_2.

Consistency and efficiency considerations

We shall need $[\partial E\{T_R(b)\}/\partial b]_{b=\beta}$ and therefore have to express $E\{T_R(b)\}$ in terms of F. Writing

$$\text{Rank}(Y_j - bx_j) = 1 + \sum_{i \neq j} U_{ij},$$

where $U_{ij} = 1$, if $Y_j - bx_j > Y_i - bx_i$, and 0 otherwise, it follows readily that

$$E\{\text{Rank}(Y_j - bx_j)\} = 1 + \sum_{i \neq j} \int F\{y - (b-\beta)(x_j - x_i)\} f(y) \mathrm{d}y,$$

since $Y_j - \alpha - \beta x_j$ and $Y_i - \alpha - \beta x_i$ are independently and identically distributed. Hence

$$[\partial E\{\text{Rank}(Y_j - bx_j)\}/\partial b]_{b=\beta} = -\sum_{i \neq j}(x_j - x_i) \int f^2(y)\mathrm{d}y = -n\bar{f}x_j \tag{5.9}$$

since $\sum x_i = 0$, where $\bar{f} = \int f^2(y)\mathrm{d}y$. Using (5.8) and (5.9), we obtain

$$[\partial E\{T_R(b)\}/\partial b]_{b=\beta} = -n\bar{f} \sum_{i=1}^{n} x_j^2 \tag{5.10}$$

To check consistency we shall assume, mainly for convenience, that $|x_j|$ is bounded, in fact $|x_i| \leqslant \frac{1}{2}$. Since $\text{var}\{\sum \text{Rank}(Y_j - bx_j)\} = 0$, we have

$$\sum_i \sum_j \text{cov}\{\text{Rank}(Y_i - bx_i), \text{Rank}(Y_j - bx_j)\} = -n\bar{v}$$

where $\bar{v}$ is the average of $\text{var}\{\text{Rank}(Y_i - bx_i)\}$, $i = 1, 2, \ldots, n$. Thus

$$\text{var}(T_R(b)) \leqslant (n/4) \max \text{var}\{\text{Rank}(Y_i - bx_i)\} + (n/4)\bar{v} \leqslant cn^3$$

where c is a constant. This result, with (5.10), can be used to establish consistency as indicated in Sections 1.3 and 1.4.

Since var $\{T_R(\beta)\} = n(n+1)\sum x_j^2/12$, the efficacy of T_R is obtained, using (5.10), as

$$e(T_R) = \bar{f}\{12\sum x_j^2\}^{1/2},$$

For comparison, the efficacy of the least-squares statistic, when F has variance $\sigma^2 < \infty$, is calculated as

$$e(T) = \{\sum x_j^2\}^{1/2}/\sigma$$

giving

$$\{e(T_R)/e(T)\}^2 = 12\bar{f}^2\sigma^2,$$

which coincides with the corresponding result for the Wilcoxon two sample rank sum statistic.

5.2.3 Sign transformation

We now consider inference about β based on a sign test of trend leading to the statistic

$$T_S(b) = \sum x_i \operatorname{sgn}[D_j(b) - \hat{D}(b)] \tag{5.11}$$

where $\hat{D}(b) = \operatorname{median}(D_j(b))$.

Hypothesis testing
For a given set of x_i values, the conditional null distribution is invariant with respect to b, since the $D_j(b)$ values are transformed to $-1, 0$, or $+1$ as indicated in (5.11). Hence, the test is unconditionally distribution-free and for fixed $x_1, x_2, \ldots, x_n$ the null distribution can be tabulated once and for all.

To test a hypothesis specifying a value of β the observed $T_S(\beta)$ is referred appropriately to the null distribution.

Example 5.9 $n = 5$: refer to the data of Example 5.1, and consider $H_0: \beta = 1$.

$$
\begin{array}{lrrrrr}
x_j: & -2 & -1 & 0 & 1 & 2 \\
y_j: & -1.35 & -1.55 & 0.48 & 1.55 & 2.21 \\
D_j(1): & 0.65 & -0.55 & -0.48 & 0.55 & 0.21 : \hat{D}(1) = 0.21 \\
\operatorname{sgn}(D_j(1) - \hat{D}(1)): & 1 & -1 & -1 & 1 & 0
\end{array}
$$

Observed $T_S(1) = 0$.

The null distribution of T_S is as follows:

$t:$	-6	-5	-4	-3	-2	-1	0	1	2	3	4	5	6
$30\Pr(T_S = t):$	1	2	2	2	3		2	6	2	3	2	2	2	1

Formulae (5.3) can be used to find $E(T_S(\beta))$ and $\mathrm{var}(T_S(\beta))$ with $\bar{d}$ and v replaced appropriately by the corresponding mean and variance calculated from the transformed $\mathrm{sgn}(D_j(\beta) - \hat{D}(\beta))$ values.

(i) n even : $n = 2k$ mean $= 0$

                                      variance $= 1$

giving

$$E(T_S(\beta)) = 0$$
$$\mathrm{var}(T_S(\beta)) = (n/(n-1))\sum x_j^2 \tag{5.12}$$

(ii) n odd : $n = 2k + 1$ mean $= 0$

$$\text{variance} = \frac{n-1}{n}$$

giving

$$E(T_S(\beta)) = 0$$
$$\mathrm{var}(T_S(\beta)) = \sum x_j^2 \tag{5.13}$$

The distribution of $T_S(\beta)$ is asymptotically normal as $n \to \infty$ with $\max x_i^2/\sum x_i^2 \to 0$, and a normal approximation for the distribution of T_S can be used for large n with formulae (5.12) or (5.13) giving the appropriate mean and variance.

Example 5.10 Take $n = 5$ and refer to Example 5.9.

$$E(T_S) = 0, \quad \mathrm{var}(T_S) = 10$$

$$\Pr(T_S \leqslant 4) \simeq \Phi\left(\frac{4.5}{\sqrt{10}}\right) = 0.92.$$

From Example 5.9 the exact probability is

$$\Pr(T_S \leqslant 4) = 27/30 = 0.9$$

Point estimation

As b varies from $-\infty$ to $+\infty$, the value of T_S changes from $+\sum|x_i|$ to $-\sum|x_i|$ in a series of steps. T_S is non-increasing in b, by an argument similar to that used to demonstrate this property for T_R. The values of b at which T_S changes are not all the pairwise slopes, but a subset of these.

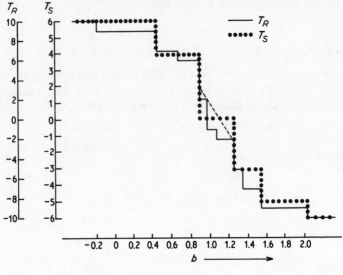

Figure 5.2

In practice it is useful to draw a graph of T_S against b. Such a graph is shown in Fig. 5.2 for the data of Example 5.9; we shall refer to it again in Example 5.11.

The solution of the estimating equation $T_S(b) = 0$, giving a point estimate of β, is readily obtained from the graph.

Confidence limits
Since the null distribution of T_S is invariant with respect to b, confidence limits can be found directly from the listing of values of b at which T_S changes and the corresponding values of T_S. We first find, from the null distribution, values t_1 and t_2 such that $\Pr(t_1 \leqslant T_S \leqslant t_2) = \gamma$. Then a $100\gamma\%$ confidence interval for β has as its lower (upper) limit the smallest (largest) value of b such that $T_S \leqslant t_2 (\geqslant t_1)$. One-sided confidence limits are found similarly.

Example 5.11 $n = 5$: refer to the data of Example 5.9. Figure 5.2 shows a graph of $T_S(b)$ against b and according to the illustrated smoothing, the point estimate of β is taken to be $b_S = 1.09$.

In the following table the 10 pairwise slopes are listed in order of magnitude, and the corresponding values of T_S and T_R are shown.

T_S	T_R		T_S	T_R
-0.2	6	10	1.07	0	-1
0.435	6	9	1.253	0	-2
0.66	4	7	1.345	-3	-5
0.89	4	6	1.55	-3	-7
0.966	0	2	2.03	-5	-9
	0	-1		-6	-10

From the null distribution of T_S, $\Pr(-5 \leqslant T_S \leqslant 5) = 28/30$, therefore a 93.3% confidence interval for β is

$$(0.435, 2.03)$$

For comparison, the graph of $T_R(b)$ against b is superimposed on the graph of $T_S(b)$ against b in Fig. 5.2. The point estimate from this graph is $b_R = 1.00$. Based on T_R, a 91.6% confidence interval for β is

$$(0.435, 1.55)$$

Consistency and efficiency considerations

The difference $D_j(b) - \hat{D}_j(b)$ can be expressed as

$$Y_j - \alpha - bx_j - (\hat{D}_j(b) - \alpha)$$

and we write $z_{0.5}(b)$ for $\hat{D}_j(b) - \alpha$, so that $z_{0.5}(b)$ satisfies

$$H_n(z_{0.5}(b)) = \tfrac{1}{2}$$

where

$$H_n(z) = \frac{1}{n} \# (Y_i - \alpha - bx_i \leqslant z, i = 1, 2, \ldots, n).$$

Now

$$H(z) = E\{H_n(z)\} = \frac{1}{n} \sum_{i=1}^{n} F\{z + (b - \beta)x_i\} \qquad (5.14)$$

and $\mathrm{var}\{H_n(z)\} \leqslant 1/4n$ for every z. Thus $z_{0.5}(b)$ is a consistent estimate of $\zeta_{0.5}(b)$ which satisfies

$$H(\zeta_{0.5}(b)) = \tfrac{1}{2}$$

Let us now impose some restrictions on the x_i values as we let $n \to \infty$, in particular that the x_i are bounded. This can be achieved, for example, by supposing that all x_i lie in a finite interval as $n \to \infty$. Then the effect on $z_{0.5}(b)$ of removing $Y_i - \alpha - bx_i$ from the set $Y_j - \alpha - bx_j$, $j = 1, 2, \ldots n$, is of order $1/n$; let the resulting median be $z_{0.5}^{(i)}(b)$. Then

$$
\begin{aligned}
E[\mathrm{sgn}\{Y_i - \alpha - bx_i - z_{0.5}(b)\}] &= E[\mathrm{sgn}\{Y_i - \alpha - bx_i - z_{0.5}^{(i)}(b)\}] \\
&\quad + O(1/n) \\
&= 1 - 2F\{\zeta_{0.5}(b) + (b - \beta)x_i\} \\
&\quad + O(1/\sqrt{n})
\end{aligned}
$$

$$(5.15)$$

after taking first a conditional expectation of $\mathrm{sgn}\{Y_i - \alpha - bx_i - z_{0.5}^{(i)}(b)\}$ with $z_{0.5}^{(i)}(b)$ fixed and then an expectation with respect to $z_{0.5}^{(i)}(b)$.

Alternatively, we note that

$$
\begin{aligned}
&\Pr[\mathrm{sgn}\{Y_i - \alpha - bx_i - z_{0.5}(b)\} = 1] \\
&= \Pr[\mathrm{sgn}\{Y_i - \alpha - bx_i - z_{0.5}(b)\} = 1] \\
&\quad \pm \Pr[(Y_i - \alpha - bx_i) \in (\zeta_{0.5}(b), z_{0.5}(b)] \\
&= \Pr[\mathrm{sgn}\{Y_i - \alpha - bx_i - z_{0.5}(b)\} = 1] + O(1/\sqrt{n})
\end{aligned}
$$

Summing terms like those in (5.15) and differentiating appropriately, we have for large n,

$$
\left(\frac{\partial T_S(b)}{\partial b}\right)_{b=\beta} \simeq -2 \sum_{i=1}^{n} x_i f\{z_{0.5}(\beta)\} \{z'_{0.5}(\beta) + x_i\}
$$

To find the value of $z'_{0.5}(\beta)$, differentiate both sides of

$$
(1/n) \sum_{i=1}^{n} F\{z_{0.5}(b) + (b - \beta)x_i\} = \tfrac{1}{2}
$$

with respect to b, and put $b = \beta$, to give

$$
f(z_{0.5}(\beta)) \sum_{i=1}^{n} (z'_{0.5}(\beta) + x_i) = 0
$$

so that, assuming $f(z'_{0.5}(\beta)) \neq 0$, we find $z'_{0.5}(\beta) = -\sum x_i = 0$. Substituting in the expression above, we obtain

$$
\left(\frac{\partial T_S(b)}{\partial b}\right)_{b=\beta} = -2f(z_{0.5}(\beta)) \sum_{i=1}^{n} x_i^2
$$

$$(5.16)$$

Using arguments similar to those leading to (5.15) we can take $Y_i - \alpha - bx_i - z_{0.5}(b)$, and $Y_j - \alpha - bx_j - z_{0.5}(b)$ to be approximately independent, to give,

$$
\begin{aligned}
\mathrm{var}\{T_S(b)\} &\simeq 4 \sum_{i=1}^{n} x_i^2 F(z_{0.5}(b) \\
&\quad + (b - \beta)x_i)\{1 - F(z_{0.5}(b) + (b - \beta)x_i)\}
\end{aligned}
$$

$$(5.17)$$

Substituting $b = \beta$ in (5.17), we obtain $\text{var}\{T_S(\beta)\} = \sum x_i^2$, which for $n \to \infty$, agrees with Expressions (5.12) and (5.13). Also, for $b \neq \beta$, $\text{var}\{T_S(b)\} \leqslant \sum_{i=1}^{n} x_i^2$, and $\text{var}\{T_S(\beta)\}$ is continuous in b.

If $f(z_{0.5}(\beta)) \neq 0$, the results (5.16) and (5.17), by application of Lemmas 1.1 and 1.2, ensure consistency of estimating and testing procedures based on T_S.

The efficacy of T_S can be found immediately using the results obtained above; in fact,

$$e(T_S) = 2f(z_{0.5}(\beta))(\sum x_i^2)^{1/2} \tag{5.18}$$

giving

$$\{e(T_S)/e(T)\}^2 = 4f^2(z_{0.5}(\beta)) \tag{5.19}$$

The result (5.19) coincides with the corresponding result for the sign test in the two-sample case. This is to be expected since the two-sample problem can be looked upon as a special case of the straight-line regression problem.

Comment

Comparing the efficacies of T_R and T_S we see that, as in previous analogous comparisons, the choice between the two statistics, from the point of view of efficiency depends on the ratio of $\bar{f}$ to $f(0) = f(z_{0.5}(\beta))$, i.e. the ratio of the mean density of F to the density of F at its median.

5.2.4 *More general rank transformations*

Write $R_i(b) = \text{Rank}(Y_i - bx_i) = \text{Rank}(Y_i - \alpha - bx_i)$ and let $H(u)$ be a monotonic continuous differentiable function of u; typically $H(u)$ may be an inverse distribution function as was discussed in Chapter 2. Define the statistic $T_{HR}(b)$ by

$$T_{HR}(b) = \sum x_i H\{R_i(b)/(n+1)\}.$$

Hypothesis testing, the null distribution

Since the distinct values of $H(R_i(b)/(n+1))$ are just the transformations $H(1/(n+1))$, $H(2/(n+1)),\ldots$ of $1/(n+1)$, $2/(n+1)$, $\ldots$, the null distribution of T_{HR} is invariant with respect to β, and it can be tabulated once and for all for every fixed set of x_i values. We need only list all permutations of the ranks $1, 2, \ldots n$ and their associated transforms.

The first two moments of the null distribution are:

$$E\{T_{HR}(\beta)\} = \sum x_i \bar{H}(\beta) = \bar{H}(\beta)\sum x_i = 0$$
$$\text{var}\{T_{HR}(\beta)\} = \{nv/(n-1)\}\sum x_i^2$$

where $v = (1/n)\sum_{i=1}^{n}\{H_i(\beta) - \bar{H}(\beta)]^2$, and $H_i(\beta) = H\{R_i(\beta)/n + 1)\}$. A normal approximation may be applied to this distribution as $n \to \infty$ of max $x_1^2/\sum x_i^2 \to 0$ as $n \to \infty$ (see Chapter 1).

Example 5.12 Refer to the data and the null hypothesis of Example 5.5: test $H_0 : \beta = 1.2$. Put $H = \Phi^{-1}$, the 'inverse normal transform'.

x_i:	-3	-1	1	3
y_i:	-1.65	-1.25	0.34	4.17
$y_i - 1.2x_i$:	1.95	-0.05	-0.86	0.57
$R_i(1.2)$:	4	2	1	3
$H_i(1.2)$:	0.84	-0.25	-0.84	0.25

Observed $T_{HR}(1.2) = -2.36$.

By listing the 24 permutations of $H_i(1.2)$ values and calculating T_{HR} for each permutation, it is readily checked that the null distribution of T_{HR} is given by

$\pm\, t$:	0.18	1.18	2.18	2.36	3.18	4.36	4.54	5.54
$24\Pr(T_{HR}=t)$:	1	2	2	2	1	2	1	1

Thus $\Pr\{|T_{HR}(1.2)| \geqslant 2.36\} = 14/24$.

Confidence limits

Since the null distribution of T_{HR} is invariant with respect to b, we need to evaluate observed $T_{HR}(b)$ for various values of b and to compare these with the null distribution to find confidence limits. The procedure is essentially the same as for T_R. In fact, since the H_i values are transformed ranks, $T_{HR}(b)$ will have 'jumps' at exactly the same values of b as T_R, that is at the pairwise slopes that have been used before.

Example 5.13 (Continuation of Example 5.12; see also Example 5.8.) Following is a list of pairwise slopes at which values of $T_{HR}(b)$ changes and the values of $T_{HR}(b)$ are shown bracketed in the relevant intervals.

(5.54) 0.2 (4.36) 0.4975 (2.36) 0.795 (1.18) 0.97 (−2.36)

1.355 (−4.36) 1.915 (−5.54)

Since $\Pr\{|T_{HR}| \leqslant 4.54\} = 22/24$ a 91.7% confidence interval for β is (0.2, 1.915). This result coincides with the result in Example 5.8; generally, especially for large n, this will not be so.

Point estimation

We obtain a point estimate of β by solving $T_{HR}(b) = 0$. For most purposes a graphical solution should be adequate. Now, referring to Fig. 5.2 where a graph of $T_R(b)$ is shown for Example 5.9 ($n = 5$), it will be noted that a graph of $T_{HR}(b)$ will be very similar to that of $T_R(b)$. Re-scaling the values of $T_{HR}(b)$ so that the largest and smallest are $+10$ and -10 respectively, the graph of $T_{HR}(b)$ would have jumps at the same values as the graph of $T_R(b)$, but the sizes of the jumps will be different. Such a graph gives a point estimate $b_{HR} = 1.09$.

Consistency and efficiency

Consistency of procedures based on T_{HR} can be established by adaptation of the types of argument used for T_R in a similar context. We need to evaluate $E\{T_{HR}(b)\}$ for b near β, and shall calculate for this purpose $(\partial E\{T_{HR}(b)\}/\partial b)_{b=\beta}$.

Let

$$\bar{F}(z; b) = (1/n) \sum_{i=1}^{n} F\{z + (b - \beta)x_i\}$$

Then

$$E\{R_j(b)/(n+1)| Y_j - \alpha - bx_j = z\}$$

$$= \{1 + \sum_{i \neq j} \Pr(Y_i - \alpha - bx_i < z)\}/(n+1) \simeq \bar{F}(z, b)$$

and

$$\mathrm{var}\{R_j(b)/(n+1)| Y_j - \alpha - bx_j = z\} = O(1/n)$$

Following the argument in Section 5.2.2, we replace $R_j(b)/(n+1)$ by $\bar{F}(Y_j - \alpha - bx_j; b)$ in calculations of $E\{T_{HR}(b)\}$. Thus

$$E\{R_j(b)/(n+1)\} \simeq \frac{1}{n} \int \sum_{i=1}^{n} F\{y + (b - \beta)(x_i - x_j)\} f(y) \mathrm{d}y$$

which agrees with the more direct derivation of Section 5.2.2 leading to (5.9). Also

$$EH\{R_j(b)/(n+1)\} \simeq \int H\left[\frac{1}{n} \sum_{i=1}^{n} \{F\{y + (b - \beta)(x_i - x_j)\}\right] f(y) \mathrm{d}y$$

giving

$$[\partial EH\{R_j(b)/(n+1)\}/\partial b]_{b=\beta} = -x_j \int H'(F(y)) f^2(y) \mathrm{d}y$$

and

$$(\partial ET_{HR}(b)/\partial b)_{b=\beta} = -\left(\sum_{i=1}^{n} x_j^2\right)\int H'(F(y)f^2(y)\,\mathrm{d}y \quad (5.20)$$

More lengthy calculations of a similar nature show that $\mathrm{var}\{T_{HR}(b)\} = O(\sum x_j^2)$, so that, applying Lemmas 1.1 and 1.2, tests and estimates based on T_{HR} are consistent.

Writing

$$T_{HR}(\beta) \simeq \sum_{i=1}^{n} x_i H\{F(Y_i - \alpha - \beta x_i)\}$$

we have

$$\mathrm{var}\{T_{HR}(\beta)\} \simeq (\sum x_i^2)\,\mathrm{var}\{H(F(Y))\}$$

$$= (\sum x_i^2)\left\{\int H^2(u)\mathrm{d}n - \left(\int H(u)\mathrm{d}n\right)^2\right\} \quad (5.21)$$

Combining (5.20) and (5.21), we obtain the efficacy of T_{HR} as

$$e(T_{HR}) = \frac{(\sum x_j^2)^{1/2}\int H(F(y))f^2(y)\mathrm{d}y}{\left\{\int H^2(u)\mathrm{d}u - \left(\int H(u)\mathrm{d}u\right)^2\right\}^{1/2}}$$

5.2.5 *Optimal weights for statistics of type T*

Returning to definition (5.1) we note that in Sections 5.2.2, 5.2.3, 5.2.4, statistics of type T with $\psi(x_j) = x_j$ have been studied, and the question that needs attention is whether 'weights' other than $\psi(x_j) = x_j$ merit serious attention.

From the point of view of efficacy, the answer seems clear: they do not because $\psi(x_j) = x_j$ gives maximum efficacy. In the case of T_R this can be seen as follows. Suppose we put

$$T_R(\mathbf{w}, b) = \sum_{i=1}^{n} w_i \mathrm{Rank}(Y_i - b x_i)$$

where $\sum w_i = 0$. Then straightforward repetition of earlier steps gives

$$\mathrm{var}\{T_R(\mathbf{w}, b)\} = \{n(n+1)/12\}\sum w_i^2$$

$$\{\partial ET_R(\mathbf{w}, b)/\partial b\}_{b=\beta} = -\bar{f}n\sum w_i x_i$$

$$\mathrm{efficacy} = \bar{f}\sqrt{12}(\sum w_i x_i)/(\sum w_i^2)^{1/2}$$

The efficacy is readily shown to be maximized by $w_i = x_i$.

Similar calculations may be made for the other statistics of type T.

A simple alternative to $\psi(x_j) = x_j$ that has received some attention in the literature, more especially in the context of joint estimation of α and β, is $w_j = \text{sgn}(x_j - \hat{x})$, where $\hat{x}$ is the median of $x_1, x_2, \ldots, x_n$; reference should be made to Brown and Mood (1951) and more recent literature on the Brown–Mood approach. In the case of testing a hypothetical value of β, the relevant calculations are effectively reduced to the analysis of 2×2 contingency table in which all four marginal totals are equal.

5.2.6 Theil's statistic, Kendall's rank correlation

In this section we shall take $x_1 < x_2 < x_3 \ldots < x_n$ unless otherwise stated. Let $z_i(b) = Y_i - \alpha - bx_i$, and put $D_{ij}(b) = z_j(b) - z_i(b)$, $i, j = 1, 2, \ldots n$. The statistic proposed by Theil (1950) for inference about β is

$$T^*(b) = \sum_{i<j} \text{sgn} \{D_{ij}(b)\}$$

Since T involves the differences of $z_i(b)$ values, the value of α is immaterial and we may write

$$T^*(b) = \sum_{i<j} \text{sgn} \{Y_i - Y_j - b(x_i - x_j)\}$$

The null distribution of T^*, that is, the distribution of $T^*(\beta)$, can be obtained by the usual permutation argument. An interesting difference between the null distribution of $T^*(b)$ and of the statistics of type B in Sections 5.2.2–5.2.5 is that it does not depend on the configuration of the $x_1, \ldots, x_n$ values. Thus it can be tabulated and used whatever the $x_1, \ldots, x_n$ values are.

It is instructive to obtain the null distribution successively for $n = 2, 3$, etc. Consider $n = 2$ where we have just 2 possible permutations of the $z_i(\beta)$ values. If $z_1(\beta) < z_2(\beta)$ the value of T^* is -1, otherwise it is $+1$. Writing T_2^* for $T(\beta)$ when $n = 2$, the distribution of T_2^* is thus:

t :	-1	$+1$
$2 \Pr(T_2^* = t)$:	1	1

For $n = 3$ straightforward enumeration of the 6 possible permutations of any three numbers, which we may denote a, b, c, gives the following distribution of T_3^*:

t :	-3	-1	$+1$	$+3$
$6 \Pr(T_3^* = t)$:	1	2	2	1

The six permutations of a, b, c can also be regarded as two permutations of a and b, in three sets with c located in positions 1, 2, 3, thus:

$$
\begin{array}{ccccccccc}
a & b & c & \quad a & c & b & \quad c & a & b \\
b & a & c & \quad b & c & a & \quad c & b & a
\end{array}
$$

In these three sets the contribution to T of the addition of the number a to the set $\{a, b\}$ is $-2, 0, +2$ if c is greater than both a and b. Therefore it is possible to represent the distribution of T_3^* as the distribution of $T_2^* + V_3$, where V_3 is independent of T_2^* and has the distribution

$$
\Pr\{V_3 = 2(i - 1)\} = 1/3, \quad i = 1, 2, 3.
$$

Proceeding in this way, one can represent T_n as

$$
T_n^* = \sum_{j=1}^{n} V_j \tag{5.22}
$$

where $V_1, V_2, \ldots, V_n$ are independent and

$$
\Pr\{V_s = s + 1 - 2j\} = 1/s, \quad j = 1, 2, \ldots, s
$$

Since we can write $V_s = s + 1 - 2U_s$ when $\Pr(U_s = j) = 1/s$ for $j = 1, 2, \ldots, s$, we see immediately that

$$
E(V_s) = s + 1 - 2(s + 1)/2 = 0
$$

and

$$
\operatorname{var}(V_s) = (s^2 - 1)/3
$$

Using (5.22) and the independence of $V_1, V_2, \ldots, V_n$, we have

$$
E(T_n^*) = 0
$$

$$
\operatorname{var}(T_n^*) = (1/3) \sum_{j=1}^{n} (j^2 - 1) = n(n - 1)(2n + 5)/18 \tag{5.23}
$$

Another advantage of the representation (5.22) is that asymptotic normality of the distribution of $T_n^*(\beta)$ follows by a simple application of Liapounov's theorem.

Reference to the well-established literature, especially Kendall (1955), will show that the statistic T^* is actually a linear function of the Kendall rank correlation coefficient, τ; in fact

$$
\tau = 2T^*/[n(n - 1)]
$$

Thus the published tables of the null distribution of τ can be used in testing a hypothesis about β when T^* is the basic test statistic.

Efficiency

Details of calculations relating to efficiency and consistency will not be given here; the efficacy of T^* can be found by steps similar to those followed for B_R.

Point estimation

Examination of the definition of $T^*(b)$ shows that $T^*(b)$ is monotonic in b and its value changes whenever b passes through one of the pairwise slope values $(Y_i - Y_j)/(x_i - x_j)$, $x_i \neq x_j$. If all x_i values are distinct, the value of T^* changes by 2 at each of these pairwise slope values, so that the solution of the estimating equation is the median of all the pairwise slopes.

To conclude this brief discussion of Theil's statistic, we note that the case where all x_i values are not distinct needs special attention because the value of T^* could be affected by the labelling of the x-values. To avoid such ambiguity, a possible definition is

$$T^*(b) = \sum_{\substack{i < j \\ x_i \neq x_j}} \text{sgn}\,[Y_i - Y_j - b(x_i - x_j)]$$

This definition implies that the null distribution depends on the number of x-values at which there are multiple Y-values, and on the multiplicities.

5.2.7 Robust transformations

The statistics considered in Sections 5.2.2–5.2.6 have all been of the 'rank' or 'sign' types that are most commonly associated with distribution-free methods. However, if any of the monotonic transformations that have become associated with robust techniques is applied to the $D_j(\beta)$ appearing in (5.1), it is again possible to develop a conditionally distribution free test procedure and a method for exact confidence limits.

Consider

$$M(b) = \sum x_i \psi(Y_i - bx_i) = \sum x_i \psi_i(b) \tag{5.24}$$

where $\rho(u)$ is monotonic continuous and differentiable almost everywhere in u. Typical examples of functions $\psi(u)$ are

$$\psi(u) = (e^u - e^{-u})/(e^u + e^{-u})$$

$$\psi(u) = \begin{cases} u, & u \leqslant k \\ k, & u \geqslant k \end{cases}$$

For a discussion of the rationale underlying the use of such transformations the reader is referred to Section 2.3.6.

Hypothesis testing

To test $H_0 : \beta = \beta_0$, the conditional null distribution of $M(\beta_0)$ has to be evaluated, and this entails, essentially, listing all the permutations of the numbers $\psi_1(\beta_0), \psi_2(\beta_0), \ldots, \psi_n(\beta_0)$. In fact, all calculations are the same as for the statistic $T(\beta_0)$ except that the numbers $y_i - \beta_0 x_i$ are replaced by $\psi(y_i - \beta_0 x_i)$, $i = 1, 2, \ldots, n$. The actual significance test is then performed by referring the observed $M(\beta_0)$ appropriately to the conditional null distribution.

Writing

$$\bar{\psi}(\beta_0) = (1/n) \sum \psi_i(\beta_0)$$

$$v_\psi = (1/n) \sum (\psi_i(\beta_0) - \bar{\psi}(\beta_0))^2$$

the mean and variance of the conditional null distribution are

$$E(M) = \bar{\psi} \sum x_j = 0$$
$$\mathrm{var}(M) = \{n v_\psi / (n - 1)\} \sum x_j^2 \tag{5.25}$$

by the steps that give (5.3). With ψ chosen such that $\mathrm{var}(\psi(Y_i))$ is finite, normal approximation of the null distribution as $n \to \infty$ will be possible if $\max x_j^2 / \sum x_j^2 \to 0$.

Confidence limits

As in the case of T, the conditional null distribution of M is not invariant with respect to b. Therefore, in order to find confidence limits the null distribution has to be evaluated for every value of b that is considered for membership of the confidence interval. We shall use the notation of Section 5.2.1 and let $N(b)$ denote the number of $M_q(b)$ values, $q \in Q'$ that are smaller than the observed $M(b)$. If $r < N(b) < n! - r$, the hypothesis $\beta = b$ is accepted and b belongs to the $100(1 - 2r/n!)\%$ confidence interval. Thus we have to examine $N(b)$ as b varies from $-\infty$ to $+\infty$. Represent $N(b)$ as

$$N(b) = \sum_{q \in Q'} I \left\{ \sum x_i \psi(y_i - bx_i) - \sum_q x_i \psi(y_{n_i} - bx_{n_i}) \right\}$$

where q refers to one of the permutations of the numbers $(y_1 - bx_1)$, $(y_2 - bx_2), \ldots, (y_n - bx_n)$, and the notation $(y_{n_i} - bx_{n_i})$ indicates that one of those permutations is used. The value of $N(b)$ will change at the solutions b_q of the equations

$$\sum x_i \{ \psi(y_i - bx_i) - \psi(y_{n_i} - bx_{n_i}) \} = 0$$

in b; if all x_i values are distinct there will be $n! - 1$ such equations.

With n large it will usually be satisfactory to use (5.25) and Fieller's theorem to obtain $100(1 - 2\gamma)\%$ confidence limits as the solution of

$$(n - 1)^{1/2} \sum x_i \psi(y_i - bx_i) = \pm u_\gamma \{ \sum (\psi_i(b) - \bar{\psi}(b))^2 \cdot (\sum x_i^2) \}^{1/2} \quad (5.26)$$

where u_γ is an appropriate normal deviate. Numerical solution of (5.26) is usually straightforward.

Point estimation

A point estimate of β is obtained by solving the estimating equation in $b : M(b) = 0$; see (5.24). In practice a numerical or graphical technique has to be employed in most cases. The estimate of β obtained from $M(b)$ is obviously related to the M-estimates of location discussed briefly in Chapter 2, and the reader is referred to that chapter for some comments on the choice of $\psi(u)$.

Efficiency

From (5.24) we have, since $Y_1, Y_2, \ldots, Y_n$ are independent, and the distribution function of $Y_i - \alpha - \beta x_i$ is $F(y)$,

$$\text{var} \{ \psi(Y_i - \beta x_i) \} = \int \psi^2(u + \alpha) f(u) \, du - \left\{ \int \psi(u + \alpha) f(u) \, du \right\}^2 = v_\psi(\alpha)$$
$$(5.27)$$

and

$$\text{var} \{ M(\beta) \} = v_\psi(\alpha) \sum x_i^2 \quad (5.28)$$

Also

$$E \{ \psi(Y_i - bx_i) \} = \int \psi \{ z + \alpha - (b - \beta) x_i \} f(z) \, dz$$

giving

$$\left\{ \frac{\partial EM(b)}{\partial b} \right\}_{b = \beta} = -(\sum x_i^2) \int \psi'(u + \alpha) f(u) \, du \quad (5.29)$$

Thus the efficacy of M is

$$e_M(\beta) = \{ v_\psi(\alpha) \}^{-1/2} \left\{ \int \psi'(u + \alpha) f(u) \, du \right\} \{ \sum x_i^2 \}^{1/2} \quad (5.30)$$

The efficacy of M depends on α. In fact, inspection of (5.24) shows that in general the point estimate of β derived from (5.24) will not be invariant with respect to α if we rewrite the estimating equation as

$$\sum x_i \psi(Y_i - \alpha - bx_i) = 0$$

The problem can be overcome by applying the robust transformation to differences $D_j(b) - \hat{D}(b)$ used in Equation (5.11) giving

$$M^*(b) = \sum x_i \psi \{D_i(b) - \hat{D}(b)\}.$$

5.2.8 An example with moderately large n

We conclude this section dealing with inference about β only with an example where $n = 21$. The main reason for its inclusion is to draw attention to the fact that, while computations involving ranking can be very tedious, practically satisfactory results can be obtained by drawing graphs of the functions T, T_R, T_S, etc., using relatively few values of b for plotting positions. Preparing programs for calculating these functions is relatively straightforward, and with $n = 21$ is feasible even with some desk calculators.

The data are given in the following table, these were generated with parameter values $\alpha = 0, \beta = 1$, and F a Cauchy distribution function $F(y) = (1/\pi) \tan^{-1}(y/\sigma) + \frac{1}{2}$ with $\sigma = 0.6745$, giving the same interquartile range as a standard normal distribution.

x	y	x	y
-10	-11.822	1	2.189
-9	-10.509	2	1.784
-8	-48.964	3	3.076
-7	-5.751	4	3.065
-6	-6.185	5	4.815
-5	-6.039	6	5.234
-4	-4.784	7	7.027
-3	-3.250	8	8.057
-2	8.034	9	9.255
-1	-1.278	10	9.194
0	1.128		

In order to find point estimates and confidence limits based on T_R and T_S, it is relatively easy to find, by trial and error, a suitable plotting range of T_R or T_S against b; we need only evaluate T_R or T_S at a selection of b-values so that the range $E(T_R) \pm u_\gamma \text{s.d.}(T_R)$ is covered, $E(T_R) = 0$ and $\text{s.d.}(T_R)$ being the mean and standard deviation of the null distribution. Following is a table of some values of T_R and T_S at selected values of b.

b	T_R	T_S	$7T_S$
0.95	393	62	434
1.00	251	38	266
1.025	134	19	133
1.05	31	-1	-7
1.10	-186	-18	-126
1.125	-325	-61	-427
1.15	-394	-82	-574

Graphs of T_R and T_S are shown in Fig. 5.3; the values of T_S have been

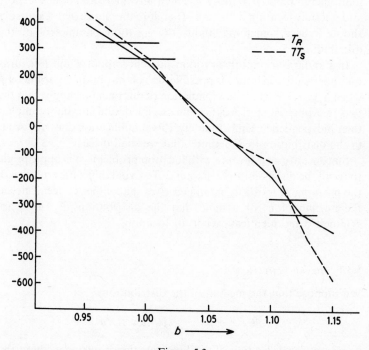

Figure 5.3

multiplied by the factor $(\max T_R / \max T_S) = 770/110$ for ready comparison. Horizontal lines are drawn at ± 1.645 s.d.$(T_R) = \pm 1.645$ $(35\sqrt{22}) = \pm 270.25$ and at ± 1.645 s.d.$(T_S) = \pm 46.97$, the latter multiplied by the factor 7 for plotting. From the graph we have the following results:

Statistic	Point estimate	90% (approx.) confidence limits
T_R	1.057	(0.993, 1.128)
T_S	1.048	(0.980, 1.111)

5.3 Joint inference about α and β

In Section 5.1 we have remarked that inference about β only can be regarded as tantamount to inference about relative location shift of several populations, and in such inference we need not be concerned with the actual measure of location. However, inference about α, or joint inference about α and β, is essentially inference about location, and not only location difference. Therefore the choice of test statistics should reflect something about the location parameters of the distributions F_i.

In standard normal theory joint inference about α and β it turns out that inference about α is possible, free of the nuisance parameter β when $\sum x_i = 0$. Part of the explanation of this phenomenon is that the two relevant test statistics are uncorrelated, and normality implies their independence. Unfortunately, this simplification does not occur in the distribution-free methods that we shall discuss.

Since we are now dealing with location problems, it is appropriate to recall the discussions of Chapter 2. To avoid difficulties to do with the meaning of location parameters, we shall either concentrate on the median of F_j or assume that the distributions F_j of Y_j are symmetric and identical except for location.

5.3.1 *Median regression*

We suppose that the median of the distribution F_j is

$$\theta_j = \alpha + \beta x_j, \qquad j = 1, 2, \ldots, n$$

A natural generalization of the sign statistic for inference about the median is the pair of statistics

$$S_1(a,b) = \sum_{j=1}^{n} \text{sgn}(Y_j - \alpha - bx_j) \left.\vphantom{\sum_{j=1}^{n}}\right\}$$

$$S_2(a,b) = \sum_{j=1}^{n} x_j\text{sgn}(Y_j - a - bx_j) \quad\quad (5.31)$$

More generally we could use

$$S_r(a,b) = \sum_{j=1}^{n} \psi_r(x_j)\text{sgn}(Y_j - a - bx_j), \qquad r = 1,2$$

but we shall see that the special choices of ψ_1 and ψ_2 giving (5.31) have optimal properties. Obviously (5.31) is inspired by the 'normal' equations arising in application of the method of least squares.

In principle it is easy to enumerate the joint null distribution of S_1 and S_2 because the random variables $\text{sgn}(Y_j - a - \beta x_j), j = 1, 2, \ldots, n$, are independent and assume the values -1 and $+1$ with equal probabilities. Straightforward calculations give

$$E\{S_1(\alpha, \beta)\} = E\{S_2(\alpha, \beta)\} = 0$$

$$\text{var}\{S_1(\alpha, \beta)\} = n$$

$$\text{var}\{S_2(\alpha, \beta)\} = \sum x_i^2 \left.\vphantom{\sum}\right\} \quad\quad (5.32)$$

$$\text{cov}\{S_1(\alpha, \beta), S_2(\alpha, \beta)\} = \sum x_i = 0 \quad \text{(by assumption)}$$

Under conditions similar to those stated in the earlier one-parameter cases for the sign statistics, the joint distribution of S_1 and S_2 tends to a bivariate normal distribution as n increases.

Example 5.14

x_i :	-3	-2	0	1	4
y_i :	0.67	2.04	2.8	3.27	3.00
$y_i - 2 - 0.4x_i$:	-0.13	0.84	0.63	0.87	-0.60

The joint null distribution of S_1 and S_2 is obtained by straightforward enumeration and it is as shown in the table below; the entries in the cells are the relevant probabilities multiplied by $2^n = 32$.

		S_1						
		-5	-3	-1	$+1$	$+3$	$+5$	
	-10			1	1			2
	-8				1	1		2
	-6		1	1				2
	-4		1	2	1			4
S_2	-2			1	2	1		4
	0	1*	1			1	1*	4
	2		1	2	1			4
	4			1	2	1		4
	6				1	1		2
	8		1	1				2
	10			1	1			2
		1	5	10	10	5	1	32

If $\alpha = 2.0$ and $\beta = 0.4$, the observed values of S_1 and S_2 are $+1$ and -2 respectively.

Inspection of the tabulation of the joint null distribution of S_1 and S_2 in Example 5.14 shows that both the marginal distributions of S_1 and S_2 are symmetric about 0. This symmetry holds in general since both S_1 and S_2 are sums of independent symmetric random variables.

The example also illustrates a difficulty which we have not encountered in the one-parameter problems that have so far been discussed. Suppose we were to consider testing the null hypothesis $\alpha = 2.0$ and $\beta = 0.4$. Then, as indicated in the example, the observed values of S_1 and S_2 are $+1$ and -2, and both of these are, relative to the standard deviations of S_1 and S_2, close to their expected values. However, there remains the question of an appropriate critical region in the (S_1, S_2)-sample space. A possible approach is outlined below: see also Chapter 1, Section 1.7.

Testing the hypothesis $H_0 : \alpha = \alpha_0, \beta = \beta_0$
We shall consider only an alternative $H_1 : \alpha \neq \alpha_0, \beta \neq \beta_0$, and use as a guide to an appropriate critical region the approach that might be used if S_1 and S_2 have an exactly joint normal distribution. Put

$$Q(a, b) = \frac{S_1^2(a, b)}{n} + \frac{S_2^2(a, b)}{\sum x_i^2} \qquad (5.33)$$

Then large values of $Q(\alpha_0, \beta_0)$ reflect observed values of $S_1(\alpha_0, \beta_0)$ and $S_2(\alpha_0, \beta_0)$ that are 'far' from their expected values. Thus $Q(\alpha_0, \beta_0)$ may be regarded as a suitable statistic for testing H_0; the critical region which it defines is the set of (S_1, S_2)-values outside a certain ellipse centred at $(0, 0)$. If the test procedure is

reject H_0 if $Q(\alpha_0, \beta_0) > C$

then the exact size of the critical region can be determined from the exact joint distribution of S_1 and S_2 after C has been specified. It can be varied by changing C. When n is sufficiently large for a normal approximation to the joint distribution of S_1 and S_2 to apply, the value of C can be found by referring to a table of the χ_2^2 distribution. For large n, the distribution of Q is approximately χ_2^2, but it will be noted that the expectation of Q is exactly 2.

Example 5.15 (Continuing Example 5.14.) Suppose we want to test $H_0 : \alpha = 2.0, \beta = 0.4$ at the 10% level. From tables, $\chi_2^2(0.90) = 4.605$, and we have $n = 5$, $\sum x_i^2 = 30$. The (S_1, S_2)-values that lie outside the ellipse

$$S_1^2/5 + S_2^2/30 = 4.605$$

are marked with a * in the tabulation of the joint distribution of S_1 S_1 and S_2 in Example 5.14. Thus $\Pr(Q > 4.605) = 2/32$ and at level $100(1 - 2/32)\%$, H_0 would only be rejected if $S_1 \pm 5$ and $S_2 = 0$.

If we put $C = 3.93 - \varepsilon$ where ε is small and positive we find that the S_1, S_2 points in the critical region are $(-5, 0), (+5, 0), (-3, 8), (+3, 8)$, giving a test size $4/32$.

Confidence regions
The arguments used for setting confidence limits in the one-parameter case carry over to the two parameter case but it must be

noted that they produce a joint confidence region for α and β. Now, as we have seen in Section 5.2, it is possible to make exact inferences about β free from the nuisance parameter α. However the roles of α and β cannot be reversed; it appears that without possibly severe loss of efficiency it is generally not possible to make exact inferences about α. This matter is taken up again in Section 5.3.2.

Thus, while we can set an exact joint confidence region for α and β, and an exact confidence interval for β alone, we cannot, generally, find an exact confidence interval for α. However, by using the point estimate of α and its standard deviation it is possible to find an approximate confidence interval for α. We shall give more details about this when dealing with point estimation.

The remarks concerning inference about α apply equally to any ordinate $\alpha + \beta x$, for by a change of x origin any such ordinate can be made the intercept of the line.

Point estimation
Since $E\{S_1(\alpha, \beta)\} = E\{S_2(\alpha, \beta)\} = 0$, a natural procedure for estimating α and β is to solve for a and b the estimating equations

$$S_1(a, b) = 0$$
$$S_2(a, b) = 0 \tag{5.34}$$

If for any b the value of a is chosen so that $S_1(a, b) = 0$, then the resulting $S_2(a, b)$ is the same as the statistic $T_S(b)$ discussed in Section 5.2, so that apart from the minor question of uniqueness that arises from the fact that $S_2(a, b)$ is, for fixed a, a step-function in b, the existence of a solution is not in question.

According to the discussion of Chapter 1, Section 1.4, the point estimates $\hat{a}$ and $\hat{b}$ of α and β will be consistent as $n \to \infty$ if for small $\Delta\alpha$ and $\Delta\beta$

$$E\{S_1(\alpha + \Delta\alpha, \beta)\} = A_{1n}\Delta\alpha$$
$$\text{Var}\{S_1(\alpha + \Delta\alpha, \beta)\} = B_{1n}$$

with $A_{1n}/B_{1n}^{1/2} \to \infty$ as $n \to \infty$, a similar condition holding for S_2.
Now

$$E\{S_1(\alpha + \Delta\alpha, \beta)\} = -n + \sum_{i=1}^{n} 2\Pr\{Y_i - (\alpha + \Delta\alpha) - \beta x_i > 0\}$$

$$\simeq -2\Delta\alpha \sum_{i=1}^{n} f_i(0), \tag{5.35}$$

and, similarly,

$$E\{S_2(\alpha, \beta + \Delta\beta)\} \simeq -2\Delta\beta \sum x_i^2 f_i(0) \qquad (5.36)$$

Also

$$\mathrm{var}\{S_1(\alpha + \Delta\alpha, \beta)\} \leqslant n$$

$$\mathrm{var}\{S_2(\alpha, \beta + \Delta\beta)\} \leqslant \sum x_i^2$$

Thus, if we consider the case where all F_i are identical except for location, then $f(0) \neq 0$ and $\sum x_i^2 \to \infty$ as $n + \infty$ ensure the consistency of the point estimation.

Writing $E_{r\alpha} = (\partial E\{S_r(a, b)\}/\partial a)_{a = \alpha, b = \beta}$, $E_{r\beta} = (\partial E\{S_r(a, b)\}/\partial b)_{a = \alpha, b = \beta}$, $r = 1, 2$, the covariance matrix of $\hat{a}$ and $\hat{b}$ can be obtained from the following approximation:

$$\begin{pmatrix} \mathrm{var}(S_1) & \mathrm{cov}(S_1 S_2) \\ \mathrm{cov}(S_1, S_2) & \mathrm{var}(S_2) \end{pmatrix}$$

$$\simeq \begin{pmatrix} E_{1\alpha} & E_{1\beta} \\ E_{2\alpha} & E_{2\beta} \end{pmatrix} \begin{pmatrix} \mathrm{var}(\hat{a}) & \mathrm{var}(\hat{a}, \hat{b}) \\ \mathrm{cov}(\hat{a}, \hat{b}) & \mathrm{cov}(\hat{b}) \end{pmatrix} \begin{pmatrix} E_{1\alpha} & E_{2\alpha} \\ E_{1\beta} & E_{2\beta} \end{pmatrix} \qquad (5.37)$$

The values of $E_{1\alpha}$ and $E_{2\beta}$ are obtainable from (5.35) and (5.36) while $E_{2\alpha}$ and $E_{1\beta}$ are obtainable similarly. With $\sum x_i = 0$ we have a rather simple result if $f_i(0) = f(0)$, $i = 1, \ldots, n$, because $E_{1\beta} = E_{2\beta} = f(0)\sum x_i = 0$, while $E_{1\alpha} = -nf(0)$, $E_{1\beta} = -f(0)\sum x_i^2$. Substituting in (5.37) we have

$$\begin{pmatrix} \mathrm{var}(\hat{a}) & \mathrm{cov}(\hat{a}, \hat{b}) \\ \mathrm{cov}(\hat{a}, \hat{b}) & \mathrm{var}(\hat{b}) \end{pmatrix} = \frac{1}{4f^2(0)} \begin{pmatrix} 1/n & 0 \\ 0 & 1/\sum x^2 \end{pmatrix} \qquad (5.38)$$

Comparing the result (5.38) with the corresponding result for least-squares estimation giving estimates $\hat{a}$ and $\hat{\beta}$, we obtain

$$\frac{\mathrm{var}(\hat{a})}{\mathrm{var}(\hat{\alpha})} = \frac{\mathrm{var}(\hat{b})}{\mathrm{var}(\hat{\beta})} = \frac{1}{\sigma^2 f^2(0)}$$

and these ratios coincide with the ratio of the variance of estimate of β based on T_S to $\mathrm{var}(\hat{\beta})$. The value $1/(\sigma^2 f(0))$ is also the relative efficiency of the mean and median in one-sample location estimation.

The approximate variances given by (5.38) can be used for setting confidence limits for α or β. However, a major problem with such a procedure is that $f(0)$ must be estimated. An estimation procedure for $f(0)$ is as follows. We use the exact confidence limit method for β to find, say, a 95% confidence interval for β. Suppose that it is

$$\{b_2(0.975), \quad b_4(0.975)\}$$

Then approximately

$$b_4(0.975) - b_2(0.975) = 2(1.96)/\{2f(0)(\textstyle\sum x_i^2)^{1/2}\} \qquad (5.39)$$

from which we obtain

$$\frac{\{b_4(0.975) - b_2(0.975)\}}{1.96}(\textstyle\sum x_i^2)^{1/2} \simeq \frac{1}{f(0)}$$

This type of calculation could be repeated for different values of the confidence coefficient, and the results suitably averaged to give an estimate of $1/f(0)$.

Efficiency

When considering the efficiency of testing a hypothesis specifying values of α and β jointly, we are led to examine the non-centrality parameter of the statistic $Q(\alpha, \beta)$ which has, asymptotically, a χ_2^2 distribution under the null hypothesis; see Section 1.7 for more details of tests involving multiple parameters.

In the following calculations we confine our attention to the case where all F_i are identical except for location. We have seen that $E\{Q(\alpha, \beta)\} = 2$, hence we shall obtain an expression for $E\{Q(\alpha + \Delta\alpha, \beta + \Delta\beta)\}$, where $\Delta\alpha$ and $\Delta\beta$ are small. Following steps like those leading to (5.35) and (5.36), we obtain

$$E\{\operatorname{sgn}(Y_i - (\alpha + \Delta\alpha) - (\beta + \Delta\beta)x_i)\} \simeq -2(\Delta\alpha + x_i\Delta\beta)f(0)$$

and

$$\operatorname{var}\{\operatorname{sgn}(Y_i(\alpha + \Delta\alpha) - (\beta + \Delta\beta)x_i\} \simeq 1 - 4(\Delta\alpha + x_i\Delta\beta)^2 f^2(0)$$

from which, remembering that $\sum x_i = 0$,

$$E\{S_1^2(\alpha + \Delta\alpha, \beta + \Delta\beta)\} \simeq n - 4f^2(0)\textstyle\sum(\Delta\alpha + x_i\Delta\beta)^2 + 4n^2\Delta^2\alpha f^2(0) \qquad (5.40)$$

It is convenient to interpret $\Delta\alpha$ and $\Delta\beta$ small as meaning $\Delta\alpha = A/\sqrt{n}, \Delta\beta = B/\sqrt{n}$, so that

$$E\{S_1^2(\alpha + \Delta\alpha, \beta + \Delta\beta)\} \simeq n - 4f^2(0)A^2 - 4f^2(0)B^2(\textstyle\sum x_i^2/n)$$
$$+ 4nf^2(0)A^2 \qquad (5.41)$$

Similarly,

$$E\{S_2^2(\alpha + \Delta\alpha, \beta + \Delta\beta)\} \simeq \textstyle\sum x_i^2 - 4f^2(0)A^2(\textstyle\sum x_i^2/n)$$
$$- 2f^2(0)AB(\textstyle\sum x_i^2/n)$$
$$- 4f^2(0)B^2(\textstyle\sum x_i^4/n)$$
$$+ 4f^2(0)B^2(\textstyle\sum x_i^2)^2/n \qquad (5.42)$$

We shall now assume that, as n increases, $\sum x_i^2/n \to K_2, \sum x_i^3/n \to K_3, \sum x_i^4/n \to K_4$, where K_2, K_3, K_4 are finite limits; this is quite a reasonable assumption and will be valid, for example, if all x_i values are constrained to lie in a fixed interval. Then, since by earlier assumption $\sum x_i^2 \to \infty$ as $n \to \infty$, we can neglect all but the first and last terms in both (5.41) and (5.42), giving

$$E\{S_1^2(\alpha + \Delta\alpha, \beta + \Delta\beta)\} \to n + 4nf^2(0)A^2$$
$$E\{S_2^2(\alpha + \Delta\alpha, \beta + \Delta\beta)\} \to K_2 + 4nf^2(0)B^2K_2^2$$

so that

$$E\{Q(\alpha + \Delta\alpha, \beta + \Delta\beta)\} \to 2 + 4f^2(0)(A^2 + B^2K_2) \qquad (5.43)$$

To compare this result with the corresponding result for the method of least squares, we assume that every Y_i has variance σ^2 and find $E\{Q_A^*(\alpha + \Delta\alpha, \beta + \Delta\beta)\}$, where

$$Q_A^*(a,b) = A_1^2(a,b)/(n\sigma^2) + A_2^2(a,b)/(\sigma^2\sum x_i^2)$$

and

$$A_1(a,b) = \sum(Y_i - a - bx_i)$$
$$A_2(a,b) = \sum x_i(Y_i - a - bx_i) \qquad (5.44)$$

The result is

$$E\{Q_A^*(\alpha + \Delta\alpha, \beta + \Delta\beta)\} \to 2 + (1/\sigma^2)(A^2 + B^2K_2)$$

Thus, taking as a measure of relative efficiency the ratio of the non-centrality parameters, we obtain the result $1/[4\sigma^2f^2(0)]$; this ratio coincides with the ARE of the mean and median in the one-sample location problem. The results can also be checked by the simpler approach indicated in Chapter 1, Section 1.7.

5.3.2 Symmetric identical F_j: Untransformed residuals

Since we are now dealing with inference about location, not merely location shift, the discussion of Chapter 2 is relevant, and in the distribution-free setting we are led to assume:

   (i) that all F_j are identical except for location;
   (ii) that F_j is symmetric.

The statistics $A_1(a,b)$ and $A_2(a,b)$ are the natural generalizations of $A(\mathbf{x}, t)$ of Section 2.3.2.
We rewrite them

$$A_1(a,b) = \sum \text{sgn}(Y_i - a - bx_i)|Y_i - a - bx_i|$$
$$A_2(a,b) = \sum x_i \text{sgn}(Y_i - a - bx_i)|Y_i - a - bx_i| \qquad (5.45)$$

In this form we see that the conditional joint null distribution of $A_1(\alpha, \beta)$ and $A_2(\alpha, \beta)$ could be derived by conditioning on the magnitudes $|Y_i - \alpha - \beta x_i|, i = 1, 2, \ldots, n$. The joint distribution can, in principle, be enumerated by listing the 2^n possible sign combinations for the magnitudes $|Y_i - \alpha - \beta x_i|, i = 1, 2, \ldots n$. We also have the following conditional parameters:

$$\text{var}\{A_1(\alpha, \beta)| \, |Y_i - \alpha - \beta x_i|, i = 1, \ldots, n\} = \sum (Y_i - \alpha - \beta x_i)^2$$
$$\text{var}\{A_2(\alpha, \beta)| \, |Y_i - \alpha - \beta x_i|, i = 1, \ldots, n\} = \sum x_i^2 (Y_i - \alpha - \beta x_i)^2$$
$$\text{cov}\{A_1(\alpha, \beta), A_2(\alpha, \beta)| \, |Y_i - \alpha - \beta x_i|, i = 1, \ldots, n\}$$
$$= \sum x_i (Y_i - \alpha - \beta x_i)^2$$

(5.46)

Under suitable conditions the joint distribution of A_1 and A_2 will be approximately normal as n becomes large. The conditions are that $\max x_i^2 / \sum x_i^2 \to 0$ as $n \to \infty$ and $\max (Y_i - \alpha - \beta x_i)^2 / \sum (Y_i - \alpha - \beta x_i)^2 \overset{P}{\to} 0$.

Testing $H_0 : \alpha = \alpha_0, \beta = \beta_0$
Although the exact joint distribution of A_1 and A_2 is known we are in the same position as before, namely that the choice of critical region, or equivalently, test statistic, is not obvious. In the previous sections we choose to use test statistics having exact expectations 2 and asymptotically χ_2^2 distributions. We follow the same procedure here.

Since the covariance of A_1 and A_2 is not zero, an appropriate test statistic is

$$Q(a, b) = (A_1(a, b), A_2(a, b)) C^{-1} (A_1(a, b), A_2(a, b))^{\mathrm{T}} \quad (5.47)$$

where

$$C = \begin{bmatrix} \sum (Y_i - a - bx_i)^2 & \sum x_i (Y_i - a - bx_i)^2 \\ \sum x_i (Y_i - a - bx_i)^2 & \sum x_i^2 (Y_i - a - bx_i)^2 \end{bmatrix}$$

With $a = \alpha, b = \beta$, we have $E\{Q(\alpha, \beta)\} = 2$ and the distribution of $Q(\alpha, \beta)$ is approximately χ_2^2 for large n.

To test $\alpha = \alpha_0, \beta = \beta_0$, the exact conditional distribution of $Q(\alpha_0, \beta_0)$ can be enumerated for small n and the observed value of $Q(\alpha_0, \beta_0)$ can be referred to it; with large n the χ^2 approximation may be used.

Testing of a hypothesis specifying α and β jointly will not be pursued in great depth because it does not seem to be as important a practical problem as testing a value of β only, or α only.

Confidence limits

By the usual argument of inverting the hypothesis-testing procedure an exact joint confidence region for α and β can be derived from the exact testing procedure outlined above. However, in practice one is more likely to be interested in a confidence interval for β or a confidence interval for α. In some practical problems one needs confidence intervals for ordinates $\alpha + \beta x$ for a succession of x-values; graphically these are usually depicted as confidence bands. Attention is drawn to the distinction between confidence limits for individual ordinates and a confidence band for the whole line $y = \alpha + \beta x$ derived from a joint confidence region for α and β; a careful discussion of this distinction in the case of normal least-squares regression is given by Kerrich (1955).

Although the value of A_1 (A_2) is invariant with respect to $b(a)$ all elements of the matrix C depend on both a and b, consequently exact inference about α, free from the nuisance parameter β is not possible; reference to Section 5.2 shows that inference about β free from α is possible.

For α one has to resort to approximate methods: one of them is to use the estimated standard error of the estimate of α. Alternatively, one could use the exact conditional distribution of A_1 that would apply if β were known and replace β by the estimate $\hat{\beta}$. Approximating this exact distribution by a normal distribution leads to solving

$$\sum(Y_i - a - \beta x_i) = \pm u_\gamma \{\sum(Y_i - a - \beta x_i)^2\}^{1/2} \qquad (5.48)$$

for a after substituting $\hat{\beta}$ for β. The result is

$$a = \hat{\alpha} \pm u_\gamma \frac{s}{\sqrt{n}} \left(\frac{n-2}{n}\right)^{1/2} (1 - u_\gamma^2/n)^{-1/2} \qquad (5.49)$$

which is similar to the result (5.6) for β.

However, we emphasize that, apart from the use of the normal approximation which is a convenience rather than a necessity, the limits (5.49) are not exact, unlike those given by (5.6) for β. The limits in (5.49) are actually estimates of the exact confidence limits. By introducing the confidence limits for β one could give confidence limits for the exact confidence limits, thus producing somewhat conservative confidence limits for α.

Point estimation

The point estimates of α and β derived by solving the estimating equations

$$A_\gamma(a,b) = 0, \qquad r = 1,2, \qquad (5.50)$$

are just the usual least-squares estimates

$$\hat{\alpha} = \sum Y_i/n, \qquad \hat{\beta} = \sum Y_i x_i / \sum x_i^2 \qquad (5.51)$$

These estimates are well known to be consistent if $\text{var}(Y_i) < \infty$ and $\sum x_i^2 \to \infty$ as $n \to \infty$.

The conditional variances of $\hat{\alpha}$ and $\hat{\beta}$ are

$$\text{var}\{\hat{\alpha} \,|\, |Y_i - \alpha - \beta x|, i = 1,2,\ldots,n\} = \sum(Y_i - \alpha - \beta x_i)^2/n^2$$

$$\text{var}\{\hat{\beta} \,|\, |Y_i - \alpha - \beta x_i|, i = 1,2,\ldots,n\} = \sum x_i^2(Y_i - \alpha - \beta x_i)^2/(\sum x_i^2)^2$$

Taking expectations, the usual results are obtained for the unconditional variances:

$$\text{var}(\hat{\alpha}) = \sigma^2/n, \qquad \text{var}(\hat{\beta}) = \sigma^2/\sum x_i^2$$

Assuming approximate normality of the conditional distributions of $\hat{\alpha}$, an approximate confidence interval for α is obtainable by using the estimated conditional variance, giving

$$\hat{\alpha} \pm u_\gamma(s/\sqrt{n})\{(n-2)/n\}^{1/2} \qquad (5.52)$$

For large n, (5.52) and (5.49) are nearly identical but (5.49) seems more satisfactory since the factor $K_\gamma = u_\gamma((n-2)/n)^{1/2}(1 - u/n)^{-1/2}$ is greater than u_γ for $n \geq 5$ and is close to the appropriate $t_{n-2}(\gamma)$ which is applicable when F is normal, as the following table shows.

	$\gamma = 0.05$		$\gamma = 0.025$	
n	K_γ	$t_{n-2}(\gamma)$	K_γ	$t_{n-2}(\gamma)$
5	1.881	2.353	3.154	3.182
10	1.723	1.860	2.234	2.306
20	1.678	1.734	2.069	2.101
∞	1.645	1.645	1.960	1.960

Efficiency

Since the inferential procedures based on A_1 and A_2 are clearly very similar to those based on the method of least squares, the classical results for that method may be expected to hold. In the case of hypothesis testing some manipulation is required to see that the classical results do apply, but since the point estimates $\hat{\alpha}$ and $\hat{\beta}$ are just the usual least-squares estimates, the classical results for estimation do apply. Briefly, they are that $\hat{\alpha}$ and $\hat{\beta}$ are the minimum-variance unbiased estimates of α and β among linear functions of $Y_1, \ldots, Y_n$ if

$\text{var}(Y_i) = \sigma^2 < \infty$; in the case of normal F they are the minimum-variance unbiased estimators.

It is also well known that the method of least squares is inconsistent if $\text{var}(Y_i)$ is not finite. For this reason it is advisable to consider the use of transformations of the $|Y_i - a - bx_i|$ to define new statistics, in a manner analogous to the use of transformations of the $|X_i - t|$ values in Chapter 2, in the one-sample location problem.

Example 5.16

x_i:	-5	-4	-3	-2	-1	0	1	2	3	4	5
y_i:	-4.79	-4.93	-2.84	-3.94	0.58	0.09	2.44	0.64	3.43	3.88	7.05

(a) S_1, S_2

Figure 5.4 shows a graph of S_2 plotted against b; the value of a is chosen to satisfy $S_1 = 0$ so that S_2 is actually identical to T_S of Section 5.2.3. Inspection of a plot of the data shows that $\hat{\beta}_S$ is near 1.0, hence T_S was calculated only for pairwise slope values near 1 and such that its absolute value is not greater than 2 s.d.(T_S), approximately. Note that s.d.$(T_S(\beta)) = $ s.d.$(S_2(\alpha, \beta)) = (\sum x_i^2)^{1/2} = \sqrt{110}$.

From this graph the point estimate of β is $\hat{\beta}_S = 1.045$. The graph also shows horizontal lines drawn at ± 1.645 s.d.$(S_2) = \pm 17.3$. The values of b at which these lines interest the graph of S_2 gives 90% confidence limits for β as 0.96 and 1.331.

Figure 5.5 gives a similar graph of $S_1(\alpha, \beta_S)$ against a and it gives as

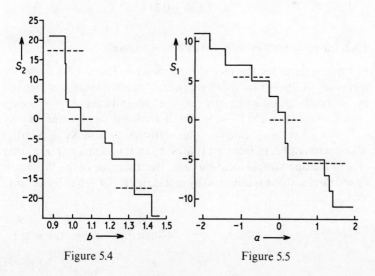

Figure 5.4   Figure 5.5

point estimate of α the median of the $Y_i - \hat{\beta}_S x_i$ values, namely, 0.295. Using s.d. $(S_1) = \sqrt{11}$, the lines drawn at $\pm 1.645\sqrt{11}$ give estimated 90% confidence limits for α as -0.75 and 1.395. Note that the limits for β are exact, apart from the use, for convenience, of the normal approximation, whereas the limits for α are estimated 90% limits.

(b) A_1, A_2
The point estimates are

$$\hat{\alpha} = 0.130$$
$$\hat{\beta} = 1.130$$

Using (5.6) to set a confidence interval for β, and using (5.49) for α, with $u_\gamma = 1.645$ we obtain the following 90% confidence intervals:

$$\alpha : 0.130 \pm 0.633$$
$$\beta : 1.130 \pm 0.213$$

Note that the interval for β is exact (apart from the use of the normal approximation for the distribution of T defined by (5.2)), whereas the interval for α is an estimated interval as described above.

The data in this example were actually generated with normally distributed residuals. The confidence intervals given by the standard normal theory are

$$\alpha : 0.130 \pm 0.633$$
$$\beta : 1.130 \pm 0.214$$

5.3.3 Identical symmetric F_j: signed-rank method

For many purposes the method discussed in Section 5.3.2 may be regarded as the primitive form which many others are derived by suitable transformations of the magnitudes $|Y_i - a - bx_i|$, $i = 1, 2, \ldots n$. Thus if $|Y_i - a - bx_i|$ is replaced in A_1, and A_2 by $\psi(|Y_i - a - bx|)$ with $\psi(n) = 1$, the statistics S_1 and S_2 of Section 5.3.1 are obtained. In fact that role of A_1 and A_2 is analogous to that of A in the one-sample location case. The analogue of the Wilcoxon signed rank statistic is obtained by replacing $|Y_i - \alpha - bx_i|$ by its rank, leading to

$$W_r(a, b) = \sum_{i=1}^{n} x_i^{r-1} \operatorname{sgn}(Y_i - a - bx_i) \operatorname{Rank}(|Y_i - a - bx_i|), r = 1, 2$$

$$(5.53)$$

These statistics and some extensions are discussed in Adichie (1967).

The conditional joint null distribution of W_1 and W_2 is, in principle, easily tabulated. Although a rank transformation has been used, this conditional distribution does depend on the sample configuration. Let $R_i = \text{Rank}(|Y_i - \alpha - \beta x_i|)$, $i = 1, 2, \ldots, n$. Then we have for this conditional joint distribution:

$$E\{W_r | C\} = 0, r = 1, 2$$
$$\text{var}\{W_1 | C\} = \sum R_i^2, \text{var}\{W_1 | C\} = \sum x_i^2 R_i^2$$
$$\text{cov}\{W_1, W_2 | C\} = \sum x_i R_i^2$$

where $W_r = W_r(\alpha, \beta), r = 1, 2$, and C denotes the sample configuration; both $\text{var}\{W_2(\alpha, \beta)\}$ and $\text{cov}\{W_1(\alpha, \beta), W_2(\alpha, \beta)\}$ depend on the sample configuration.

For the unconditional joint distribution the covariance matrix of W_1 and W_2 is obtained quite easily from the results given above; they give:

$$E(W_r) = 0, r = 1, 2$$
$$\text{var}(W) = n(n + 1)(2n + 1)/6$$
$$\text{var}(W_1) = (n + 1)(2n + 1)(\sum x_i^2)/6$$
$$\text{cov}(W_1, W_2) = 0, \quad \text{if} \quad \sum x_i = 0$$

It is possible to show that the joint distribution of W_1 and W_2 is asymptotically normal, with suitable conditions on the x_i-values.

Testing $H_0 : \alpha = \alpha_0, \beta = \beta_0$

An exact test of a joint hypothesis specifying α and β is possible since the exact joint null distribution of W_1 and W_2 is known. As test statistic we could use a Q statistic based on the exact conditional joint distribution of W_1 and W_2, or more simply,

$$Q_W(\alpha_0, \beta_0) = 6\{W_1^2(\alpha_0, \beta_0) + W_2^2(\alpha_0, \beta_0)/\sum x_i^2\}/\{(n + 1)(2n + 1)\}$$

whose exact distribution can be tabulated. It can be approximated by a χ_2^2-distribution under suitable conditions.

Other hypothesis tests

Composite hypotheses where only one of α or β is specified, for example, $H_0 : \beta = \beta_0$, cannot be tested exactly using W_1 and W_2. The other parameter is a nuisance parameter so the problem cannot be solved exactly. An approximate test is possible, analogous to the finding of estimated confidence limits, as we shall explain below.

Point estimation

Point estimates a_W and b_W are found as the solutions of

$$W_r(a, b) = 0, \quad r = 1, 2 \qquad (5.54)$$

For any selected b, the solution $a(b)$ of $W_1(a, b) = 0$ is the median of the pairwise averages of the intercepts $y_i - bx_i$; this follows from the analogy between W_1 and the Wilcoxon signed rank statistic of Chapter 2 and its associated Hodges–Lehmann point estimate. The point estimate of β can then be found graphically by plotting $W_2(a(b), b)$ against b. By considering small changes in b it can be shown that this function is non-increasing in b. Let the solution of $W_2(a_W(b), b)$ be b_W; then the point estimate of α is taken to be $a(b_W) = a_W$.

Confidence limits

In principle a joint confidence region for α and β based on W_1 and W_2 is easily found; in practice calculations could be tedious. However, as we have remarked before, determining a joint confidence region for α and β seems to be of much less practical importance than determining confidence limits for α and β separately.

Now, we have seen that it is possible to find a point estimate of β, and an exact confidence interval, free from the nuisance parameter α, if we use the methods of Section 5.2. Thus whether W_1 and W_2 should be used to find a confidence interval for β depends on whether a significant gain in efficiency is obtained, as against the simpler procedures of Section 5.2. As we shall see below the asymptotic efficiency of estimating β by W_1 and W_2 is the same as using the simpler rank statistic T_R of Section 5.2. Therefore we shall consider only the setting of a confidence interval for α.

Suppose we know the true value of β. Then the random variables $Y_i - \beta x_i$ are identically distributed with location parameter α, and W, becomes, effectively the signed rank statistic W of Chapter 2, and the point estimate of α is the Hodges–Lehmann type of estimate calculated from observed $Y_i - \beta x_i$ values. Further, exact confidence limits for α can be found by the procedures described in Chapter 2. If β is unknown we can carry through the same calculations with β replaced by an estimate of β. In the present context we take the estimate to be b_W or b_{TR}. The formal confidence interval calculation produces an estimate of the exact confidence interval that would have been obtained had β been known.

Example 5.17   Use the data of Example 5.16. Following the method of Section 5.2.3 the point estimate of β yielded by the statistic T_R is 1.148. The values of $y_i - 1.148x_i$ are

x_i:	-5	-4	-3	-2	-1	0
$y_i - 1.148x_i$:	0.950	-0.338	0.604	-0.644	1.728	-0.090

x_i:	1	2	3	4	5
$y_i - 1.148x_i$:	1.292	-1.656	-0.014	-0.72	1.31

Now using the method of Section 2.2 the point estimate of α is the median of the pairwise averages of the $y_i - 1.148x_i$ values listed above, namely 0.126, and the associated 95% confidence limits are

$$(-0.525, \quad 0.819)$$

Efficiency

By formula (1.11) the covariance matrix of the point estimates a_W and b_W is

$$E_W^{-1} C_W \left(E_W^{\mathrm{T}} \right)^{-1}$$

where C_W is the covariance matrix of $W_0(\alpha, \beta)$ and $W_1(\alpha, \beta)$, that is

$$C_W = \begin{pmatrix} n & 0 \\ 0 & \sum x_i^2 \end{pmatrix} \frac{(n+1)(2n+1)}{6}$$

and

$$E_W = \begin{pmatrix} (\partial E W_1(a,b)/\partial a)_{a=\alpha, b=\beta} & (\partial E W_1(a,b)/\partial b)_{a=\alpha, b=\beta} \\ (\partial E W_2(a,b)/\partial a)_{a=\alpha, b=\beta} & (\partial E W_2(a,b)/\partial b)_{a=\alpha, n=\beta} \end{pmatrix}$$

Expressions for the derivatives appearing in the matrix E_W can be obtained by steps similar to those involved in the derivations represented by equations (2.48) to (2.53) in Chapter 2. The result is

$$E_W = 2n\bar{f} \begin{pmatrix} n & 0 \\ 0 & \sum x_i^2 \end{pmatrix}$$

the diagonal element of E_W is actually $2n\bar{f}\sum x_i = 0$ since $\sum x_i = 0$. Thus, as $n \to \infty$ the covariance matrix of a_n, b_n becomes

$$\frac{1}{12\bar{f}^2} \begin{pmatrix} 1/n & 0 \\ 0 & 1/\sum x_i^2 \end{pmatrix} \qquad (5.55)$$

The result (5.55) shows that, in terms of relative efficiency of estimation, the estimates a_W and b_W stand in the same relation to the method of least squares as does the Hodges–Lehmann estimator to the mean.

Estimating the standard errors of a_W and b_W
There are several ways in which estimates of s.e.(a_W) and s.e.(b_W) can be calculated. Apparently the most direct of these is to try constructing an estimate of $\bar{f}$. However, density estimation is not a trivial exercise, even in simple cases. Experience in the one-parameter case suggests that estimating the elements of E_W^{-1} might be simpler.

Write

$$E_W = \begin{pmatrix} H_{11} & H_{12} \\ H_{21} & H_{22} \end{pmatrix}$$

and note that $H_{21} = H_{12} = 0$ so that

$$E_W^{-1} = \begin{pmatrix} 1/H_{11} & 0 \\ 0 & 1/H_{22} \end{pmatrix}$$

Thus we can estimate the elements of E_W by the slopes at (a_W, b_W) of the graphs of W_1 and W_2 against a and b. More directly the elements of E_W^{-1} can be estimated by the slopes of the same graphs with a regarded as a function of W_1 and b as a function of W_2, both near W_1 and $W_2 = 0$.

5.3.4 *Identical symmetric F_j: scores based on ranks*

We consider briefly the extension of the ideas of Section 2.2 to straightline regression. Instead of transforming $y_i - a - bx_i$ to Rank $(Y_i - a - bx_i)$, we put Rank $(Y_i - a - bx_i) = R_i(a, b)$ and transform to $H\{R_i(a, b)/n + 1)\}$ giving the statistics

$$W_{Hr}(a, b) = \sum x_i^{r-1} \operatorname{sgn}(Y_i - a - bx_i) H\{R_i(a, b)/(n+1)\} \quad r = 1, 2 \tag{5.56}$$

The large-sample properties of tests and estimates based on statistics of this type have been examined in detail by Adichie (1967). Essentially they are similar to those of statistics $T_H(\ )$ of Chapter 2. Typically score functions H are taken to be inverse distribution functions; if the 'right' score function is selected, statistics can be obtained that are 'best'.

5.3.5 *Identical symmetric F_j: robust transformations*

Instead of first transforming $|Y_i - a - bx_i|$ to Rank$(|Y_i - a - bx_i|)$, a transformation $(|Y_i - a - bx_i|)$ may be applied directly to the absolute deviations. Transformations ψ that have been used in 'robust' statistics are typically of the types exemplified in Section 2.2. Our main concern here is not with robustness and its ramifications, but to indicate that, conditioning on the absolute values $|Y_i - \alpha - \beta x_i|$, $i = 1, 2, \ldots, n$, it is possible to obtain the exact conditional joint distribution of

$$M_{\psi r}(\alpha, \beta) = \sum x_i^{r-1} \operatorname{sgn}(Y_i - \alpha - \beta x_i)\psi(|Y_i - \alpha - \beta x_i|), \quad r = 1, 2$$
$$(5.57)$$

The relevant arguments are exactly parallel to those of Section 5.3.2 above, and it is possible to obtain an exact joint confidence region for α and β based on M_1 and M_2. However, the question of choice of a critical region for a joint test of α and β arises, as it did in Sections 5.3.1–5.3.3. We outline a possible approach.

Conditionally on fixed $|Y_i - \alpha - \beta x_i|$, $i = 1, 2, \ldots, n$, we have

$$E\{M_\psi(\alpha, \beta)\} = 0$$
$$\operatorname{var} M_{\psi r}(\alpha, \beta) = \sum x_i^{2r} \psi_i^2, \qquad r = 1, 2$$

where $\psi_i = \psi(|Y_i - \alpha - \beta x_i|)$, and $\operatorname{cov}\{M_{\psi 1}, M_{\psi 2}\} = \sum x_i \psi_i^2$.

For testing a hypothesis specifying α and β we can define a critical region through the use of a statistic

$$Q_M(\alpha, \beta) = (M_{\psi 1}, M_{\psi 2}) V^{-1} (M_{\psi 1}, M_{\psi 2})^{\mathrm{T}}$$

where

$$V = \begin{pmatrix} \sum \psi_i^2 & \sum x_i \psi_i^2 \\ \sum x_i \psi_i^2 & \sum x_i^2 \psi_i^2 \end{pmatrix}$$

whose exact conditional distribution can be tabulated quite easily with reasonably small n; this distribution has mean exactly 2 and is approximately χ_2^2 under suitable conditions on the $|Y_i - \alpha - \beta x_i|$ values.

For efficiency calculations, and large-sample approximations, we note that, unconditionally,

$$E\{M_{\psi r}(\alpha, \beta)\} = 0$$
$$\operatorname{var}\{M_{\psi r}(\alpha, \beta)\} = (\sum x_i^{2r-2})\} \int \psi^2(|u|) f(u) \, du$$
$$\operatorname{cov}\{M_{\psi 1}, M_{\psi 2}\} = 0$$

If we define $\psi(u)$ such that $\psi(-u) = -\psi(u)$, then we can write

$$M_{\psi r}(\alpha, \beta) = \sum x_i^r \psi(Y_i - \alpha - \beta x_i), \quad r = 1, 2$$

and we have

$$E\{M_{\psi r}(\alpha, \beta)\} = \sum x_i^r \int \psi\{u + (\alpha - a) + (\beta - b)x_i\}f(u)\,\mathrm{d}u \quad r = 1, 2$$

(5.58)

Using (5.58) we obtain

$$(\partial EM_{\psi 0}/\partial a)_{a = \alpha, b = \beta} = -n \int \psi'(u)f(u)\,\mathrm{d}u$$

$$(\partial EM_{\psi 0}/\partial b)_{a = \alpha, b = \beta} = -\left\{ \int \psi'(u)f(u)\,\mathrm{d}u \right\} \sum x_i = 0$$

$$= (\partial EM_{\psi 1}/\partial a)_{a = \alpha, b = \beta}$$

$$(\partial EM_{\psi 1}/\partial b)_{a = \alpha, b = \beta} = -\left\{ \int \psi'(u)f(u)\,\mathrm{d}u \right\} \sum x_i^2$$

and using formula (1.11) the covariance matrix of the estimates (a_ψ, b_ψ) of (α, β) obtained by solving

$$M_{\psi r}(a, b) = 0, \qquad r = 0, 1$$

is

$$\begin{pmatrix} 1/n & 0 \\ 0 & 1/\sum x_i^2 \end{pmatrix} \frac{\int \psi^2(u)f(u)\,\mathrm{d}u}{\{\int \psi'(u)f(u)\,\mathrm{d}u\}^2}$$

(5.59)

In practice one may have to estimate the value of the factor $\int \psi^2(u)f(u)\,\mathrm{d}u/\{\int \psi'(u)f(u)\,\mathrm{d}u\}^2$ that appears in (5.59). One way of doing this is to use the values $\hat{u}_i = y_i - a_\psi - b_\psi x_i$ as realizations of a random variable U having density $f(u)$.
Then

$$(1/n)\sum \psi^2(u_i) \text{ estimates } \int \psi^2(u)f(u)\,\mathrm{d}u$$

and

$$(1/n)\sum \psi'(u_i) \text{ estimates } \int \psi'(u)f(u)\,\mathrm{d}u.$$

EXERCISES

5.1   Tabulate the exact null distribution of the sign statistic

$$T_S(b) = \sum x_i \operatorname{sgn}\{D_j(b) - \hat{D}(b)\},$$

defined in Section 5.2.3, for the x_i set

$$-3, -1, +1, +3.$$

5.2 Suppose that there are $3k$ equally spaced x_i values in a straight line regression problem and that the medians of the lower, middle and upper third of the x_i-values are $\hat{x}_-, \hat{x}_0, \hat{x}_+$. Let

$$R_m(b) = \hat{x}_+ \sum_{i=2k+1}^{3k} \text{Rank}\,(Y_i - bx_i) - \hat{x}_- \sum_{i=1}^{k} \text{Rank}\,(Y_i - bx_i)$$

Obtain a large k approximate formula for the variance of the point estimate of the slope β based on the statistic R_M in terms of the common 'error' distribution F.

5.3 Consider n odd $= 2k + 1$ and a scoring system whereby $Y_i - bx_i$ is transformed first to $\text{Rank}\,(Y_i - bx_i) = R_i(b)$ and then to

$$Z_i(b) = \begin{cases} [-\ln 2\{1 - R_i(b)/(n+1)\}]/\gamma + 1; & R_i > 2k+1 \\ 0; & R_i = 2k+1 \\ [\ln\{2R_i(b)/(n+1)\}]/\gamma - 1; & R_i < 2k+1 \end{cases}$$

(*Note* that as $\gamma \to \infty$ $Z_i \to \text{sgn}\,(Y_i - bx_i)$.)

For the straight line regression data below find 90% (approximately) two sided confidence limits for the slope β using Z values first with $\gamma = 100$, then $\gamma = 1$, and the test statistic $B_Z(b) = \sum x_i Z_i(b)$.

x_i:	-5	-4	-3	-2	-1	0	1	2	3	4	5
y_i:	4.8	6.4	8.6	2.8	8.4	13.2	8.1	11.7	13.9	18.8	16.8

5.4 Suppose that Y_i observed at x_i has distribution function $F_i(y; \beta, x_i)$ with median βx_i, $i = 1, 2, \ldots, n$, and that $Y_1, Y_2, \ldots Y_n$ are independent. Let

$$S(b) = \sum x_i \, \text{sgn}\,(Y_i - bx_i).$$

Express $E\{S(b)\}$ in terms of $F_i(y; \beta, x_i)$, $i = 1, 2, \ldots n$.

5.5 Refer to Exercise 5.4 and let

$$F_i(y; \beta, x_i) = \begin{cases} 1 - \exp(-y \ln 2/(\beta x_i)), & y \geq 0 \\ 0 & y < 0. \end{cases}$$

Find the large n variance of the point estimate of β obtained by solving $S(b) = E(S(\beta))$, for b. Assume that as n increases the x_i values remain approximately equally spaced between 0 and ξ.

Compare your result with the variance of the maximum likelihood estimate of β.

5.6 Using the data of Exercise 5.3 and the rank statistic

$$B_R(b) = \sum x_i \, \text{Rank} \, (Y_i - bx_i)$$

obtain a point estimate of the slope β in the straight line regression.

5.7 One way of calculating an estimate of the intercept parameter α in straight line regression is to obtain, first, an estimate $\hat{\beta}$ of β, for example, as in Exercise 5.6. Then the differences $Y_i - \hat{\beta}x_i$ $i = 1, 2, \ldots n$ are treated as if they are independently distributed with location parameter α, and one of the one-sample techniques is used to estimate α.

Obtain the Hodges–Lehmann estimate of α using the differences $Y_i - \hat{\beta}x_i, i = 1, 2, \ldots n$.

5.8 The Brown–Mood (1955) procedure for inference in straight line regression can be considered as follows. Let $\hat{x}$ denote the median of the x_i values. Replace x_i in the sign statistic $T_s(b)$, Equation 5.11, by $\text{sgn}(x_i - \hat{x})$, $i = 1, 2, \ldots n$.

Obtain the efficacy of the resulting statistic and compare it with the efficacy of T_S.

5.9 The data given below on the cost of operating a hosiery mill and its output are quoted in Mansfield (1980) p. 360 (*Statistics for Business and Economics*, Norton, New York)

| x : Output (tons) | 1 2 4 5 6 7 8 8 9 |
| y : Cost (thousands of dollars) | 2 3 4 5 6 6 7 8 8 |

Obtain an estimate of b by the method of Exercise 5.8.

5.10 Let b_m denote the Brown–Mood estimate of β. A corresponding estimate of α is the median, a_m of the differences $Y_i - b_m x_i$, $i = 1, 2, \ldots n$.

Obtain a_m for the data in Exercise 5.9.

CHAPTER 6

Multiple regression and general linear models

6.1 Introduction

In the 'parametric' literature on the topics of this chapter it has become part of the folklore that problems of the types considered in Chapters 4 and 5 are merely special cases of the 'general linear hypothesis' whose treatment is well established and known. With the treatment of the general model being so accessible one may argue that treatment of the special cases is relatively uninteresting and needs to be examined in detail only for individual applications. To a certain extent this argument can be defended, but it must be remembered that much of the apparent simplicity of the treatment of the general linear model is associated with the common assumption of identically normally distributed errors.

It is certainly possible to develop a distribution-free approach to a general linear model along lines somewhat similar to those of the traditional treatment. But it soon becomes clear that exact inferential statements can only be made about rather uninteresting questions, unless one considers generalizations with certain special properties which can perhaps be labelled properties of orthogonality.

We shall, therefore, consider separately generalizations of the topics of earlier chapters in different directions. Thus we shall generalize straight-line regression to 'plane regression' with $k = 2$ independent variables; then to multiple linear regression with $k > 2$; the two-sample problem to the k-sample problem with $k > 2$; and so on.

6.2 Plane regression: two independent variables

Let $Y_1, Y_2, \ldots, Y_n$ be independent random variables, identically distributed except for location; thus we take the distribution function of Y_j to be $F(y - \alpha - \beta_1 x_{1j} - \beta_2 x_{2j})$, $j = 1, 2, \ldots, n$. We have linear regression of Y on two independent non-stochastic variables x_1 and x_2. The coordinates (x_{1j}, x_{2j}) of the points in the (x_1, x_2) plane at which Y is observed are usefully summarized by the design matrix

$$\mathbf{X} = \begin{bmatrix} 1 & x_{11} & x_{21} \\ 1 & x_{12} & x_{22} \\ \vdots & & \\ 1 & x_{1n} & x_{2n} \end{bmatrix}$$

The natural generalization of the statistic A given in Chapter 5 is the pair of statistics

$$\begin{aligned} A_1(\mathbf{Y}; \beta_1, \beta_2) &= \sum x_{1j}(Y_j - \beta_1 x_{1j} - \beta_2 x_{2j}) \\ A_2(\mathbf{Y}; \beta_1, \beta_2) &= \sum x_{2j}(Y_j - \beta_1 x_{1j} - \beta_2 x_{2j}) \end{aligned} \tag{6.1}$$

The choice of these statistics is inspired by the 'normal equations' of the method of least squares. They do not depend on α, and are suitable for inference about β_1 and β_2 only. Following the approach developed in earlier chapters, we shall consider the distributions of A_1 and A_2 and various modifications of them under certain permutation schemes, and use these for inferential procedures.

6.2.1 Inference about $\boldsymbol{\beta} = (\beta_1, \beta_2)$: statistics A_1 and A_2

Different permutation schemes are possible, the most direct of these being based on the fact that all differences $D_j = Y_j - \beta_1 x_{1j} - \beta_2 x_{2j}$ are identically distributed. Thus by taking all permutations of the observed $d_j = y_j - \beta_1 x_{1j} - \beta_2 x_{2j}$ to be equally probable, we can tabulate a joint conditional distribution of the statistics A_1 and A_2 defined in (6.1).

We shall assume throughout that $\sum x_{1j} = \sum x_{2j} = 0$. Further, although it may not always be possible to arrange this without a rotation of the x-axes which may be regarded as artificial, it will sometimes also be assumed that $\sum x_{1j} x_{2j} = 0$.

Example 6.1 This example is used to illustrate the permutation scheme mentioned above. We have $n = 4$ observations as follows:

y_j	x_{1j}	x_{2j}	$y_j - x_{1j} - 2x_{2j}$
-2.312	-1	-1	-0.312
2.156	-1	1	0.156
-0.140	1	-1	-0.140
3.612	1	1	-0.388

The last column in the table above is obtained with $\beta_1 = 1, \beta_2 = 2$, giving observed

$$A_1(\mathbf{y}_j; 1, 2) = -0.372$$
$$A_2(\mathbf{y}_j; 1, 2) = 0.220$$

There are 24 arrangements of the d_j values; the following table gives three of them with the corresponding values of A_1 and A_2.

x_{1j}	x_{2j}	Permutations of d_j values		
-1	-1	-0.388	-0.388	-0.388
-1	1	-0.312	-0.312	-0.140
1	-1	-0.140	0.156	-0.312
1	1	0.156	-0.140	0.156
	A_1	0.716	0.716	0.372
	A_2	0.372	-0.220	0.716 etc.

The joint conditional permutation distribution of A_1 and A_2 is as follows; the table entries are probabilities multiplied by 24.

		A_2					
		-0.716	-0.372	-0.220	0.220	0.372	0.716
	-0.716	0	1	1	1	1	0
	-0.372	1	0	1	1	0	1
A_1	-0.220	1	1	0	0	1	1
	0.372	1	0	1	1	0	1
	0.716	0	1	1	1	1	0

Let $\bar{d} = \sum d_j/n$ and $\sigma^2 = \sum(d_j - \bar{d})^2/n$. Then

$$E(A_r) = \bar{d} \sum x_{rj} = 0, r = 1, 2, \text{ and}$$

$$\mathrm{var}(A_1) = \sigma^2 \sum x_{1j}^2 + \left(\sum_{j \neq k} x_{1j}x_{1k} \right)[-\sigma^2/(n-1)]$$

$$= \sigma^2 \left(\frac{n}{n-1} \right) \sum x_{1j}^2, \tag{6.2}$$

since $\sum x_{ij} = 0$. Similarly,

$$\operatorname{var}(A_2) = \sigma^2 \left(\frac{n}{n-1} \right) \sum x_{2j}^2$$

$$\operatorname{cov}(A_1, A_2) = \sigma^2 \left(\frac{n}{n-1} \right) \sum x_{1j} x_{2j} \tag{6.3}$$

Thus, since $\sum x_{1j} x_{2j} = 0, A_1$ and A_2 are uncorrelated. However, reference to Example 6.1 shows that A_1 and A_2 are not, in general, independent.

Another point to note about the conditional distribution when $\sum x_{1j} x_{2j} = 0$ is that although the value of observed A_1 is invariant with respect to β_2, its joint distribution with A_2, and its marginal distribution, are not. Thus β_2 acts as a nuisance parameter for inference about β_1, and it appears to be impossible to free β_1 from β_2, at least with the permutation scheme that we are now considering.

Exact joint inference about β_1 and β_2 is possible through the use of the joint conditional permutation distribution of A_1 and A_2. However, as we have seen before, there is some arbitrariness in the choice of a test statistic based on both A_1 and A_2.

Testing $H_0: \beta_1 = \beta_1^\circ, \beta_2 = \beta_2^\circ$

The first step is to tabulate the joint conditional distribution of A_1 and A_2, using the values β_1° and β_2° for β_1 and β_2. We accept H_0 if the observed $\{A_1(\mathbf{y}; \beta_1^\circ, \beta_2^\circ), A_2(\mathbf{y}; \beta_1^\circ, \beta_2^\circ)\}$ is not an extreme point. By convention we judge whether an observed point is extreme in terms of a measure of distance from its expectation with reference to the joint distribution of A_1 and A_2.

With $\sum x_{1j} x_{2j} = 0$ we take as measure of distance, that is as a test statistic,

$$Q(\mathbf{y}; \beta_1^\circ, \beta_2^\circ) = \frac{1}{\sigma^2} \left\{ \frac{A^2(\mathbf{y}; \beta_1^\circ, \beta_2^\circ)}{\sum x_{1j}^2} + \frac{A^2(\mathbf{y}; \beta_1^\circ, \beta_2^\circ)}{\sum x_{2j}^2} \right\} \left(\frac{n-1}{n} \right) \tag{6.4}$$

whose exact distribution can be obtained from the joint distribution of A_1 and A_2.

Example 6.2 Suppose we use the data of Example 6.1 to test H_0: $\boldsymbol{\beta} = (1, 2)$ against $H_1: \boldsymbol{\beta} \neq (1, 2)$.

$$\text{Observed } Q = 0.8011,$$

and $\Pr(Q > \text{observed } Q) = 16/24$; H_0 is accepted.

When n is large the joint distribution of A_1 and A_2 is, under suitable conditions on the design matrix X, approximately normal. The expectation of Q is exactly 2, and when the joint distribution of A_1 and A_2 is approximately normal, its distribution is approximately χ_2^2

Point estimation and confidence limits
Point estimates b_1 and b_2 of β_1 and β_2 are obtained by solving the estimating equations

$$A_r(\mathbf{y}_1; b_1, b_2) = 0; \quad r = 1, 2 \qquad (6.5)$$

These point estimates are the usual 'least-squares' estimates and their efficiency properties are well known.

An exact joint confidence region for β_1 and β_2 can be obtained via the hypothesis test based on Q. For large n, the distribution of Q is approximately χ_2^2, so an approximate confidence region for (β_1, β_2) is given by

$$\frac{\left\{\sum x_{1j}(y_j - \beta_1 x_{1j} - \beta_2 x_{2j})\right\}^2}{\sum x_{1j}^2} + \frac{\left\{\sum x_{2j}(y_j - \beta_1 x_{1j} - \beta_2 x_{2j})\right\}^2}{\sum x_{2j}^2}$$

$$\leqslant \frac{C}{(n-1)} \sum (y_j - \bar{y} - \beta_1 x_{1j} - \beta_2 x_{2j})^2 \qquad (6.6)$$

where C is a selected quantile of the χ_2^2 distribution. Note that the χ^2 approximation is a mathematical convenience so the region is exact in the sense of Section 1.2. Fixing β_2, it is easily checked that in the quadratic in β_1, defined by using the equality in (6.6) and taking all terms to the left side, the coefficient of β_1 is positive. Thus the region defined by (6.6) is closed with a roughly elliptical shape.

6.2.2 Inference about β_1 or β_2 individually using A_1 and A_2: restricted randomization

We now consider specializing the design matrix in order to be able to make exact inferences about β_1 and β_2 separately. We begin by arranging the (x_{1j}, x_{2j}) points in a rectangular lattice pattern. It is somewhat more convenient in this case to use a double subscript for the y values; thus y_{rs} corresponds to the point

$$(x_{1r}, x_{2s}), \quad r = 1, 2, \dots, p, s = 1, 2, \dots, q$$

with $pq = n$; see Fig. 6.1. We take as before $\sum_r x_{1r} = \sum_s x_{2s} = 0$, which in this case ensures $\sum_{rs} x_{1r} x_{2s} = 0$. Let us now focus attention on the

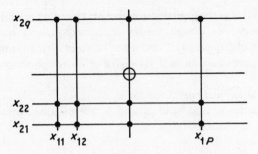

Figure 6.1

collection of observations at $x_2 = x_{21}$; there are p of them. Since α and $\beta_2 x_{21}$ are fixed quantities for this set of observations, we can use a statistic

$$A_{11} = \sum x_{1r}(y_{r1} - \beta_1 x_{1r}) = \sum x_{1r} D_{r1}(\beta_1)$$

for inference about β_1. The permutation procedure we use here is permutation of the D_{1r} values within the 'row' defined by $x_2 = x_{21}$. In short, at the fixed value $x_2 = x_{21}$ the problem, as far as β_1 is concerned, is reduced to one of straight-line regression, and we use the methods of Chapter 5.

The restricted, within-row randomization can be carried out to evaluate the distributions of statistics

$$A_{1s} = \sum x_{1r}(y_{rs} - \beta_1 x_{1r}) = \sum x_{1r} D_{rs}(\beta_1), \qquad s = 1, 2, \ldots q$$

and naturally one defines the statistic

$$A_{1.} = \sum_{s=1}^{q} A_{1s} = \sum_{s=1}^{q} \sum_{r=1}^{p} x_{1r}(y_{rs} - \beta_1 x_{1r})$$

for inference about β_1.

With $\sum x_{1j} x_{2j} = 0$ in A_1 defined by (6.1), we see that A_1 and $A_{1.}$ are identical. However, the randomization schemes under which the exact conditional distributions of these two statistics are evaluated are quite different. Although the randomization method within each row is the same as that exploited in straight-line regression, and the principles for hypothesis testing and finding confidence limits are the same, the details are somewhat more complicated because the distribution of $A_{1.}$ is the distribution of the sum of independent random variables, each of whose distribution is obtained by randomization. Example 6.3 gives an illustration of the tabulation of the distribution of $A_{1.}$.

Hypothesis test of β_1 only

Consider $H_0 : \beta_1 = \beta_1^\circ$. The first step in evaluating the conditional null distribution of $A_{1.}$ is to calculate the values of $D_{rs}(\beta_1^\circ)$. Then the $p!$ permutations of these $D_{rs}(\beta_1^\circ)$ values within each row are listed giving $p!$ possible valus for each A_{1s}, $s = 1, 2, \ldots, q$. The exact distribution of $A_{1.}$ then has $(p!)^q$ 'points' of equal probability.

To perform the test of H_0, the observed value of $A_{1.}$ is referred suitably to the null distribution of $A_{1.}$.

The enumeration of the null distribution described above is clearly a large task, even for only moderately large p and q. Let

$$D_{.s}(\beta_1^\circ) = (1/p) \sum_{r=1}^{p} D_{rs}(\beta_1^\circ)$$

and

$$\sigma_{1s}^2(\beta_1^\circ) = (1/p) \sum_{r=1}^{p} [D_{rs}(\beta_1^\circ) - D_{.s}(\beta_1^\circ)]^2, \qquad s = 1, 2, \ldots, q$$

Then using the results in Chapter 5, the conditional distribution of A_{1s} gives

$$E(A_{1a}) = 0$$

$$\text{var}(A_{1s}) = \left(\frac{p}{p-1}\right) \sigma_{1s}^2(\beta_1^\circ) \sum_{r=1}^{p} x_{1r}^2$$

Since we randomize independently within the separate rows we have

$$E(A_{1.}) = 0$$

$$\text{var}(A_{1.}) = \left(\frac{p}{p-1}\right)\left(\sum_{r=1}^{p} x_{1r}^2\right)\left(\sum_{s=1}^{q} \sigma_{1s}^2(\beta_1^\circ)\right) \qquad (6.7)$$

Moreover, from Chapter 5 we can conclude that since $A_{1.}$ is the sum of independent random variables each of whose distribution tends to normality as p increases, the distribution of $A_{1.}$ tends to normality as p increases. On the other hand, if we keep p fixed, but increase q, we see that $A_{1.}$ is the sum of not necessarily identically distributed random variables, but with moderate constraints on the σ_{1s}^2 values the distribution of $A_{1.}$ again tends to normality.

Thus, for moderately large p or q and using a normal approximation to the conditional distribution of $A_{1.}$, the arithmetic of hypothesis testing about β_1 becomes relatively easy. Only calculation of the observed $A_{1.}$ and its variance according to (6.7) is required, the rest of the calculations being arithmetically trivial.

Confidence limits for β_1

The null distribution of $A_1.$ is used in the same way that the null distribution of A is used in Chapter 5. For any trial b_1 the crucial quantity, on which depends the decision as to whether b_1 is in the confidence region, is $N(b_1)$, the number of the $(p!)^q$ possible $A_1.$ values that are greater than or equal to the observed $A_1.$.

The following example explains the procedure.

Example 6.3 In the following table the entries are observed values of Y.

x_2 \\ x_1	-1	0	$+1$
-1	-3.072	-1.648	-0.623
$+1$	0.905	1.910	2.985

We shall find an upper confidence limit for β_1 using within-row permutation.

Suppose we take as a trial β_1 value $b_1 = 1.2$. Then the values of $y_{r1} - 1.2x_{1r}, y_{r2} - 1.2x_{1r}, r = 1, 2, 3$, are:

x_2 \\ x_1	-1	0	1
-1	-1.872	-1.648	-1.823
$+1$	2.105	1.910	1.785

giving observed

$$A_1. = A_{11} + A_{12} = +0.049 - 0.320 = -0.271$$

According to the randomization scheme there are 3! equiprobable values of each of A_{11} and A_{12}, so that listing the 36 equiprobable values of $A_1.$ is straightforward. The values are

A_{11}	:		0.049,	0.175,	0.224	
A_{12}	:		0.125,	0.145,	0.320	
$A_1.$	:	0.020,	0.029,	0.050,	0.076,	0.096,	0.099
		0.145,	0.146,	0.174	0.244,	0.271,	0.300
		0.349,	0.369,	0.370,	0.419,	0.495,	0.544

each of these numbers appearing with both a negative and positive sign. There are $N(b_1) = 27$ possible $A_{1.}$ values greater than or equal to observed $A_{1.} = -0.271$.

As b_1 increases $N(b_1)$ increases, and we take as the upper $100(32/36)\% = 89\%$ confidence limit for β_1 the value of b_1 at which $N(b_1)$ changes from 32 to 33.

In this example, jumps in the value of $N(b_1)$ occur at all pairwise slopes calculated within rows and at their averages. The value of $N(b_1)$ jumps from 32 to 33 at $b_1 = \{(-1.648 + 3.072) + (2.985 - 0.903)/2\}/2 = 1.2320$. We take this value to be an upper 89% confidence limit for b_1.

The data in this example were generated with normally distributed residuals. The classical procedure gives an upper 89% confidence limit of 1.242; interpolation in one of the standard tables was used to give $t_3(0.39) = 1.54$.

With larger values of p and q we can use the normal approximation described above. For example, two-sided 90% confidence limits for β_1 are obtained as the set of b_1 values satisfying

$$\left\{ \sum_{s=1}^{q} \sum_{r=1}^{p} x_{1r}(y_{rs} - b_1 x_{1r}) \right\}$$

$$\leqslant 1.645^2 \left(\frac{p}{p-1} \right) \left(\sum_{r=1}^{p} x_{1r}^2 \right) \left(\sum_{s=1}^{q} \sigma_{1s}^2(b_1) \right) \quad (6.8)$$

Using the equality in (6.6) produces a quadratic expression in b_1 which can be simplified; it is left as an exercise.

Inference about linear combinations of β_1 and β_2

The reason for considering the restricted randomization schemes of this section, which are only really feasible with X suitably designed, is that it enables one to make exact inferences about β_1 or β_2 individually; as we have noted in Section 6.2.1, with an arbitrary X this is not generally possible.

A similar problem arises in connection with inference about $\omega_1 \beta_1 + \omega_2 \beta_2$, where ω_1 and ω_2 are constants; in fact β_1 and β_2 are just special cases of such linear combinations. Now, considering $\beta_1^* = \omega_1 \beta_1 + \omega_2 \beta_2$ as the parameter of interest amounts to reparametrizing the original problem. Suppose that $\omega_1^2 + \omega_2^2 = 1$, which entails no loss of generality, so that we can put $\omega_1 = \sin \theta, \omega_2 = \cos \theta$. Then the plane

$$\eta = \alpha + \beta_1 x_1 + \beta_2 x_2$$

becomes

$$\eta = \alpha + (\beta_1 \sin \theta + \beta_2 \cos \theta)x_1^* + (\beta_2 \sin \theta - \beta_1 \cos \theta)x_2^*$$
$$= \alpha + \beta_1^* x_1^* + \beta_2^* x_2^*$$

where

$$\begin{pmatrix} x_1 \\ x_2 \end{pmatrix} = \begin{pmatrix} \sin \theta & -\cos \theta \\ \cos \theta & \sin \theta \end{pmatrix} \begin{pmatrix} x_1^* \\ x_2^* \end{pmatrix}$$

The new (x_1^*, x_2^*) axes are obtained by rotating the old axes through an angle θ.

Exact inference about β_1^* will now be possible if we have observations at sets of points in the (x_1^*, x_2^*) plane that lie on straight lines parallel to the x_1^* axis. However, if the original design was such as to give a rectangular grid of observation points, a rotation of axes may yield points that satisfy our criterion, but each set will not necessarily contain the same number of points. The following example illustrates some of these remarks.

Example 6.4 Suppose that the (x_1, x_2) points are regularly spaced on a rectangular lattice with $x_{1r} = -2, -1, 0, 1, 2$, and $x_{2s} = -2, -1, 0, 1, 2$, so that $n = 25$.

Suppose we put $\beta_1^* = \dfrac{1}{\sqrt{2}}\beta_1 + \dfrac{1}{\sqrt{2}}\beta_2$ then the new axes are as shown in Fig. 6.2.

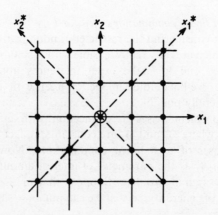

Figure 6.2

Referring to the diagram we see that there are several sets of lines parallel to the x_1^* axis containing observational points. Randomization along each line leads to exact inference about β_1^*. These methods are exactly like those described for β_1 except that the number of points in each line is not the same; the required alterations in formulae are straightforward.

6.2.3 Rank statistics: inference about β

We shall follow the methods of Section 6.2.1 except that the statistics A_1 and A_2 are replaced by

$$R_i(Y_j;\beta) = \sum_{j=1}^{n} x_{ij} \operatorname{Rank}(Y_j - \beta_1 x_{1j} - \beta_2 x_{2j}), \qquad i = 1, 2 \quad (6.9)$$

Under overall randomization the joint conditional distribution of R_1 and R_2 is readily tabulated by the method shown in Example 6.1. Since the collection of ranks is invariant with respect to the observations, the distribution so tabulated is also the unconditional distribution.

Example 6.5 Use the design matrix with $n = 4$ of Example 6.1. The joint distribution of R_1 and R_2 is as follows:

R_1 \\ R_2	-4	-2	0	2	4
-4		1	2	1	
-2	1		2		1
0	2	2		2	2
2	1		2		1
4		1	2	1	

The entries in the table above are probabilities multiplied by 24.

The variance and covariance formulae (6.2) and (6.3) apply with $\sigma^2 = (n^2 - 1)/12$, so that

$$\operatorname{var}(R_i) = \frac{n(n+1)}{12} \sum_{j=1}^{n} x_{ij}^2, \quad i = 1, 2$$

$$\operatorname{cov}(R_1, R_2) = \frac{n(n+1)}{12} \sum x_{1j} x_{2j}$$

The joint distribution of R_1 and R_2 tends to normality as n increases if $\max(x_{ij}^2)/\sum x_{ij}^2 \to 0$, $i = 1, 2$.

Using the joint distributions described above we can proceed with joint inference about β_1 and β_2. The problems associated with inference about β_1 and β_2 individually arise again and we shall look at these separately.

Testing $H_0: \beta_1 = \beta_1^\circ, \beta_2 = \beta_2^\circ$

As with A_1, A_2, we have to decide upon a test statistic, and we use the quadratic form on R_1 and R_2 that has expectation 2 under H_0. If we assume $\sum x_{1j}x_{2j} = 0$, we put

$$Q_R(\mathbf{Y}, \boldsymbol{\beta}^\circ) = \beta\frac{12}{n(n+1)}[\{R_1(\mathbf{Y};\boldsymbol{\beta}^\circ)\}^2/\sum x_{1j}^2 + \{R_2(\mathbf{Y};\boldsymbol{\beta}^\circ)\}^2/\sum x_{2j}^2]$$

Example 6.6 Data on the conversion percentage in an experiment on caustic fusion (Data of F. R. Lloyd reproduced in Burr (1974), p. 405).

Time (hours) x_1	Average fusion temperature (°C) x_2	Conversion percentage y	Rank (y)
3 (-1)	297.5 (-1)	62.7	1
3 (-1)	322.5 (0)	76.2	3
3 (-1)	347.5 (1)	80.8	5
6 (0)	297.5 (-1)	80.8	6
6 (0)	322.5 (0)	89.2	8
6 (0)	347.5 (1)	78.6	4
9 $(+1)$	297.5 (-1)	90.1	9
9 $(+1)$	322.5 (0)	88.0	7
9 $(+1)$	347.5 (1)	76.1	2

A test of $\beta_1 = 0$, $\beta_2 = 0$ is required.

The x_1 and x_2 scales are by simple transformations converted to the values shown in brackets in the table above; the ranks of the y-values are shown with the tie at 80.8 broken arbitrarily. Now we obtain

$$\sum x_{1j}^2 = \sum x_{2j}^2 = 6, \ \sum x_{1j}x_{2j} = 0$$
$$R_1 = 9 + 7 + 2 - (1 + 3 + 5) = 9$$
$$R_2 = 2 + 4 + 5 - (1 + 6 + 9) = -5$$
$$Q_R = 2.36.$$

The observed value $Q_R = 2.36$ is clearly not significant compared with χ_2^2.

Joint confidence regions for β_1 and β_2

Using Q_R and the usual inverse argument, it is in principle straightforward to find a joint confidence region for β_1 and β_2. Generally the confidence region determined in this way has an irregular outline when n is small, and the calculations needed to determine it are tedious.

When n becomes large, with suitable and practically reasonable restrictions on $\mathbf{X}$, both R_1 and R_2 treated as functions of β_1 and β_2 for fixed $\mathbf{y}$ can be approximated by linear functions over a small region of $\boldsymbol{\beta}$ values such that

$$Q_R(\mathbf{y}, \boldsymbol{\beta}) \leqslant \chi_2^2(1 - \alpha) \tag{6.10}$$

Consequently the outline of the joint confidence region is elliptical and it can be sketched by finding relatively few points on its boundary.

Computationally the setting of a confidence region can be performed by selecting a sequence of β_1 values and finding the smallest and largest β_2 values that satisfy (6.10). Putting $y_j^* = y_j - \beta_1 x_{1j}$, we know from Section 5.2.2 that as β_2 varies the ranks of $y_j^* - \beta_2 x_{2j}$ only change at values of β_2 given by pairwise slopes $(y_j^* - y_1^*)/(x_{2j} - x_{2i})$, hence it is relatively easy to find the critical β_2 values for any chosen value of β_1.

Point estimation
Putting

$$R_i(\mathbf{y}, \mathbf{b}) = 0 \quad \text{for } i = 1, 2 \tag{6.11}$$

gives two estimating equations whose solutions may be taken as point estimates of β_1 and β_2. Since R_1 is monotonic in b_1 for fixed b_2, the function $b_1(b_2)$, giving the value of b_1 at which $R_1 = 0$ is single-valued. A similar remark applies to $b_2(b_1)$ and R_2. Therefore, apart from having to adopt a convention to define a unique $b_1(b_2)$ for every b_2, and likewise for $b_2(b_1)$, the point estimates are unique.

6.2.4 *Restricted randomization: inference about β_1 or β_2 only based on R_1 and R_2*

In connection with the statistics A_1 and A_2 it was noted that while exact inference about β_1 and β_2 jointly is possible, it is generally impossible to make exact inferences about β_1 or β_2 or $\omega_1\beta_1 + \omega_2\beta_2$ separately, unless X is such that restricted randomization can be used. These remarks apply also to R_1 and R_2; in particular, exact inference about β_1 is possible by using the restricted randomization scheme.

We shall consider here only the case where the design matrix X is as described in Section 6.2.2; 'untidy' design matrices will be discussed in Section 6.2.8. Using a notation corresponding to that of Section 6.2.2 we put

$$R_{1s} = \sum_{r=1}^{p} x_{1r} \operatorname{Rank}(y_{rs} - \beta_1 x_{1r}), \quad s = 1, 2, \ldots, q$$

and

$$R_{1.} = \sum_{s=1}^{q} R_{1s}$$

Under the 'within-row' randomization scheme, the exact distribution of R_{1s}, and hence that of $R_{1.}$, is easily tabulated in principle. The realized value of this statistic $R_{1.}$, and its conditional distribution, are invariant with respect to β_2, hence it can be used for exact inference about β_1. The null distribution is also invariant with respect to β_1, so it can be tabulated once and for all for a given X.

We have

$$E(R_{1.}) = 0 \quad \text{if} \quad \sum x_{1j} = 0$$

and for the variance of $R_{1.}$ we can use (6.7), noting that

$$\sigma^2_{1s}(\beta_1) = (p^2 - 1)/12$$

giving

$$\operatorname{var}(R_{1.}) = \frac{qp(p+1)}{12} \sum_{r=1}^{p} x_{1r} \tag{6.12}$$

Testing $H_0 : \beta_1 = \beta_1^\circ$
The observed $R_{1.}$ is simply referred to the null distribution of $R_{1.}$ which is described above. If n is sufficiently large, a normal approximation, based on the variance (6.12) can be used.

Example 6.7 See also Example 6.6. In this case tabulation of the exact distribution of $R_{1.}$ is relatively easy and produces:

r:	0	±1	±2	±3	±4	±5	±6
$216.\Pr(R_{1.} = r)$:	24	36	27	14	12	6	1

To test $H_0 : \beta_1 = 0$ we arrange the data as follows:

x_2	x_1			
	-1	0	$+1$	R_{1s}
-1	62.7 (1)	80.8 (2)	90.1 (3)	$+2$
0	76.2 (1)	89.2 (3)	88.0 (2)	$+1$
1	80.8 (3)	78.6 (2)	76.1 (1)	-2

The ranks within rows are shown in brackets. We have $R_{1.} = +1$, clearly not significant.

Confidence limits for β_1

Since the conditional null distribution of $R_{1.}$ is invariant with respect to β_1, confidence limits can be determined using the null distribution by steps like those involving the use of T_R in Section 5.2.2.

As b_1 increases, $R_{1.} = R_{1.}(b_i)$ is a non-increasing step-function and varies between $-\bar{R}$ and $+\bar{R}$, where $\bar{R}$ depends on the x_{1r} configuration. Let r_1 and r_2 be such that

$$\Pr(r_1 \leqslant R_{1.} \leqslant r_2) = s/(p!)^q$$

and suppose that $b_1(1), b_1(2), \ldots, b_1(m)$ are the points at which jumps in $R_{1.}(b_1)$ occur. These are the pairwise slopes within rows; generally $m = qp(p-1)/2$.

Then let $b_1(m_1)$ be the point at which $R_{1.}(b_1)$ jumps from $r_2 + 1$ to r_2; this is the lower limit of the two-sided $100\,s/(p!)^q\%$ confidence interval. The upper limit is determined similarly.

Example 6.8 From Examples 6.6 and 6.7 we obtain 9 pairwise b_1 slopes within the three rows in which x_2 is fixed. These values are given below, and the values of $R_{1.}$ are shown in brackets in the relevant intervals.

$$(+6) - 2.5 \ (45) - 2.35 \ (+3) - 2.2 \ (+2) - 1.2 \ (+1)$$
$$5.9 \ (-1) \ 9.3 \ (-2) \ 13.0 \ (-3) \ 13.7 \ (-5) \ 18.1 \ (-6)$$

A $100(1 - 14/216)\% = 93.5\%$ confidence interval for β_1 is

$$(-2.35, 13.7).$$

6.2.5 Consistency and efficiency of the rank methods

Consistency

We shall discuss only the consistency of point estimates derived from

(6.11); other consistency results follow in a similar manner. In this discussion, as in dealing with consistency of T_R in Section 5.2.2, we assume $|x_{rj}| \leq \frac{1}{2}, r = 1, 2, j = 1, 2, \ldots, n$. In practice this can always be arranged by a rescaling of x. Then it is easily shown by exactly the same argument as in Section 5.2.2 that var $\{R_i(\mathbf{b})\} = C_i n^3$ as $n \to \infty$.

Also, by steps similar to those leading to (5.9),

$$\gamma_{st} = \left(\frac{\delta E\{R_s(\mathbf{b})\}}{\delta b_t}\right)_{\mathbf{b} = \beta} = \begin{cases} -n\bar{f}\sum x_{sj}^2, & s = t = 1, 2 \\ 0 & s \neq t \end{cases} \tag{6.13}$$

Therefore, for small $\Delta\beta_1$ and $\Delta\beta_2$,

$$E\{R_s(\boldsymbol{\beta} + \Delta\boldsymbol{\beta})\} \simeq -n\Delta\beta_s\bar{f}\sum_{j=1}^{n} x_{sj}^2, \qquad s = 1, 2.$$

Using the results of Section 1.3, consistency of point estimation follows.

Efficiency

We consider efficiency of estimation only. The large-n covariance matrix of $\hat{\beta}_1$ and $\hat{\beta}_2$ derived from (6.11) is C_β, satisfying

$$\begin{pmatrix} \gamma_{11} & \gamma_{12} \\ \gamma_{21} & \gamma_{22} \end{pmatrix} C_\beta \begin{pmatrix} \gamma_{11} & \gamma_{21} \\ \gamma_{12} & \gamma_{22} \end{pmatrix} \simeq \begin{pmatrix} \text{var}(R_1(\boldsymbol{\beta})) & \text{cov}(R_1(\boldsymbol{\beta}), R_2(\boldsymbol{\beta})) \\ \text{cov}(R_1(\boldsymbol{\beta}), R_2(\boldsymbol{\beta})) & \text{var}(R_2(\boldsymbol{\beta})) \end{pmatrix}$$

$$\tag{6.14}$$

and giving

$$C_\beta \simeq \frac{1}{12\bar{f}^2} \begin{pmatrix} 1/\sum x_{1j}^2 & 0 \\ 0 & 1/\sum x_{2j}^2 \end{pmatrix} \tag{6.15}$$

Thus, relative to the method of least squares the efficiency is $12\bar{f}^2\sigma^2$, a result equal to several previous ones.

Considering estimation of β_1 through R_1 the point estimate is $\hat{\beta}_1$. Using results from Chapter 5, we have

$$\left\{\frac{\partial E\{R_1(b_1)\}}{\partial b_1}\right\}_{b_1 = \beta_1} = -\bar{f}pq \sum_{r=1}^{p} x_{1r}^2$$

Also,

$$\text{var}(R_1(\beta_1)) = \frac{pq(p+1)}{12} \sum_{r=1}^{p} x_{1r}^2$$

from (6.12). Therefore, the approximate variance of $\hat{\beta}_1$ for large p and

q is

$$\operatorname{var}(\hat{\beta}_1) \simeq 1 \bigg/ \left\{ 12\bar{f}^2 q \sum_{r=1}^{p} x_{1r}^2 \right\}$$

This result, it will be noted, coincides with $\operatorname{var}(\hat{\beta}_1)$ given by (6.14) as $n \to \infty$.

6.2.6 *Sign transformations: inference about* $\boldsymbol{\beta}$

The randomization procedures of Section 6.2.1 are followed but the statistics A_1 and A_2 are modified to become

$$S_r(\boldsymbol{\beta}) = \sum x_{rj} \operatorname{sgn}(D_j(\boldsymbol{\beta}) - \hat{D}(\boldsymbol{\beta})), \qquad r = 1, 2 \tag{6.16}$$

where $\hat{D}(\beta) = \operatorname{median} \{D_1(\boldsymbol{\beta}), \ldots, D_n(\boldsymbol{\beta})\}$

$$D_j(\beta) = Y_j - \beta_1 x_{1j} - \beta_2 x_{2j}, j = 1, 2, \ldots, n$$

Tabulation of the conditional joint null distribution of S_1 and S_2 is straightforward and the moments up to order 2 are readily obtained by using (6.3) with σ^2 replaced by the appropriate values:

$$n \quad \text{even} \quad : \quad \sigma^2 = 1$$
$$n \quad \text{odd} \quad : \quad \sigma^2 = (n-1)/n$$

Thus: $E\{S_r(\boldsymbol{\beta})\} = 0, r = 1, 2$, and the covariance matrix is

$$\left(\frac{n}{n-1} \right) \begin{pmatrix} \sum x_{1j}^2 & \sum x_{1j} x_{2j} \\ \sum x_{1j} x_{2j} & \sum x_{2j}^2 \end{pmatrix} \tag{6.17}$$

for n even; for n odd, the factor $(n/(n-1))$ in (6.17) becomes 1. The joint distribution of S_1 and S_2 tends to normality as $n \to \infty$ if $\max(x_{rj}^2)/\sum x_{rj}^2 \to 0, r = 1, 2$.

For inference about $\boldsymbol{\beta}$ the statistic

$$Q_S(\beta) = \sigma^2 \{S_1^2(\boldsymbol{\beta})/\sum x_{1j}^2 + S_2^2(\boldsymbol{\beta})/\sum x_{2j}^2\} \tag{6.18}$$

may be used, with σ^2 as given above, if $\sum x_{1j} x_{2j} = 0$.

Hypothesis testing
Example 6.9 Refer to Example 6.6, and test $H_0 : \beta_1 = 0, \beta_2 = 0$. In this case the observed D_j values are just the original observations (y) shown in Example 6.6, so that $\hat{D} = 80.8$. Breaking the tie at 80.8 as was

done in Example 6.6, we obtain observed values

$$S_1 = 3, \quad S_2 = -3$$

and $\sigma^2 = 8/9$, giving

$$Q_S(\boldsymbol{\beta}) = (9/8)\{9/6 + 9/6\} = 3.375$$

Taking the distribution of Q_S to be approximately χ_2^2, we accept H_0.

Joint confidence regions and point estimation
An exact joint confidence region for β_1 and β_2 can be determined through the use of Q_S in the same way that Q_R can be used for the same purpose, as discussed in Section 6.2.3.

Point estimates of β_1 and β_2 are obtainable as solutions of the estimating equations

$$S_r(\mathbf{b}) = 0, \quad r = 1, 2 \tag{6.19}$$

Consistency and efficiency
Demonstrating consistency of inference based on S_1 and S_2 is relatively straightforward and is left as an exercise. By methods similar to those used with T_S in Chapter 5, the covariance matrix of the point estimates derived from (6.19) tends, as $n \to \infty$, to

$$\frac{1}{4f(0)^2}\begin{bmatrix} 1/\sum x_{1j}^2 & 0 \\ 0 & 1/\sum x_{2j}^2 \end{bmatrix} \tag{6.20}$$

if $\sum x_{1j}x_{2j} = 0$, where $f(0)$ is the density of F at its median.

6.2.7 *Restricted randomization using sign statistics: inference about* β_1 *or* β_2 *separately.*

Consider a design matrix as in Section 6.2.2. Let

$$\hat{D}_s(\beta_1) = \text{median}\,(y_{1s} - \beta_1 x_{11}, y_{2s} - \beta_1 x_{12}, \ldots, y_{ps} - \beta_1 x_{1p})$$
$$s = 1, 2, \ldots, q,$$

and put

$$S_{1s} = \sum_{r=1}^{p} x_{1r}\,\text{sgn}\,\{y_{rs} - \beta_1 x_{1r} - \hat{D}_s(\beta_1)\}, \qquad s = 1, 2, \ldots, q$$

and

$$S_{1.} = \sum_{s=1}^{q} S_{1s}$$

The conditional null distribution of every S_{1s} is like that of T_s, Section 5.2.3. The statistics $S_{11}, S_{12}, \ldots, S_{1q}$ are conditionally independent and hence enumeration of the distribution of $S_{1\cdot}$ is straightforward in principle. With $\sum x_{1j} = 0$ we have

$$E(S_{1\cdot}) = 0$$

and it is easily seen that

$$\mathrm{var}\,(S_{1\cdot}) = q \sum_{r=1}^{p} x_{1r}^2 \quad \text{or} \quad \frac{q(p-1)}{p} \sum_{r=1}^{p} x_{1r}^2 \tag{6.21}$$

according as p is even or odd. By arguments similar to those applied to $R_{1\cdot}$, Section 6.2.4, the distribution of $S_{1\cdot}$ tends to normality as p and $q \to \infty$ under suitable conditions on $\mathbf{X}$.

Inferential procedures are similar to those described for $R_{1\cdot}$, and we shall illustrate them in Example 6.10.

Example 6.10 The following data from a test of corrosion resistance of steel plates are reproduced from Johnson and Leone (1964, p. 442). The Y-values, three of them at every combination of time (x_1) and temperature (x_2), are weight losses (g); one of the sets of data at time 8 hrs, 180°F was omitted.

(x_2) Temperature (°F)	(x_1) Time (hours)					$S_{1s}(0)$
	(-2) 4	(-1) 6	(0) 8	(1) 10	(2) 12	
140 (-3)	0.0065	0.0057	0.0058	0.0070	0.0119	
	0.0047	0.0045	0.0055	0.0110	0.0148	14
	0.0132	0.0054	0.0064	0.0111	0.0100	
160 (-1)	0.0083	0.0065	0.0093	0.0094	0.0117	
	0.0080	0.0078	0.0084	0.0126	0.0107	15
	0.0059	0.0097	0.0105	0.0130	0.0105	
180 $(+1)$	0.0073	0.0084	0.0140	0.0169	0.0182	
	0.0094	0.0092	0.0115	0.0207	0.0299	16
	0.0068	0.0086	0.0130	0.0112	0.0356	
200 $(+3)$	0.0171	0.0142	0.0185	0.0325	0.0485	
	0.0142	0.0146	0.0160	0.0095	0.0349	10
	0.0111	0.0111	0.0176	0.0149	0.0380	

It is suggested that a plane regression of Y on x_1 and x_2 might be appropriate. In the following calculations the rescaled x_1 and x_2 values as indicated are used.

Testing $\beta_1 = 0$
The values of $S_{1s}(\theta)$ are shown with the table above, giving

$$S_{1.} = 55$$

In this example we have to make some small, obvious modifications to the formula (6.21) owing to there being three Y-values at each (x_1, x_2) combination. Thus, within rows there are $p = 15$ observations so that

$$\operatorname{var}(S_{1.}) = 4\left(\frac{14}{15}\right) \sum_{r=1}^{15} x_{1r}^2 = (4 \times 14 \times 30)/15 = 112$$

Taking $S_{1.}$ to be approximately normally distributed, the observed value of $S_{1.}$ is clearly highly significant, since $55/\sqrt{112} = 5.2$.

95% confidence interval for β_1
Figure 6.3 shows a graph of $S_{1.}(b)$ plotted against b for a few values of b. The plotting values of b were determined using the least-squares estimate to indicate a starting value.

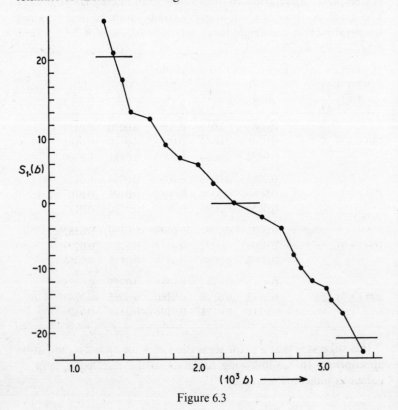

Figure 6.3

Using the normal approximation for $S_{1.}$ we note that $S_{1.}(\beta_1)$ should lie between $-1.96\sqrt{112} = -20.7$ and $+1.96\sqrt{112} = 20.7$ with probability 0.95. Thus horizontal lines drawn at these two values to intersect the graph of $S_{1.}(b)$ give the confidence interval for β_1 as (0.001 34, 0.003 26); the point estimate of β_1 is 0.002 28.

6.2.8 Grouping and restricted randomization

In previous sections we have shown how restricted randomization can be used to make inference about β_1 (or β_2) alone, free from the nuisance parameter β_2 (or β_1). Strictly such a procedure is possible whenever there are multiple $x_2(x_1)$ values corresponding to at least one $x_1(x_2)$ level. Clearly it may happen that only a small proportion of the data may be suitable for application of the restricted randomization technique, leading to unacceptable loss of efficiency for the sake of statistical exactness. In the examples discussed above, the problem did not arise because of the nature of the designs. However cases with 'untidy' design matrices do arise in practice and we shall examine a compromise method, using grouping, for dealing with them.

Suppose that n is moderately large and that we group the x_2 values into q classes. Treating the x_2 values within each class as being identical, we can then perform a restricted randomization analysis with respect to β_1. A grouping error is, of course, introduced. An idea of its size can be obtained by, for instance, comparing the estimate of β_1 calculated from the restricted randomization statistic $A_{1.}$ with the ungrouped least-squares estimate.

Apart from the grouping error there is also loss of efficiency as a result of performing a 'within-rows' analysis for β_1 without using the 'between-rows' information about β_1. A brief review of the usual least-squares procedures gives an indication of the loss of efficiency. Suppose that there are q distinct x_2 levels at which there are $n_1, n_2, \ldots, n_q$ values of x_1. Let the x_2 values be $x_{21}, x_{22}, \ldots, x_{2q}$, and denote the n_r values of x_1 at level x_{2r} by $x_{11}, \ldots, x_{1n_r}$. Suppose that

$$\sum n_s x_{2s} = 0 = \sum_{s=1}^{q} \sum_{j=1}^{n_r} x_{1j}, \text{ and put}$$

$W_1 =$ within-group sum of squared differences for x_1

$B_1(B_2) =$ between-group sum of squared differences for $x_1(x_2)$

$B_{12} =$ between-group sum of x_1, x_2 cross-products.

These sums of squares and cross-products are defined in the usual way associated with the analysis of variance.

Suppose that the Y values have variance σ^2. Then the variance of the full least-squares estimate of β_1 is $\sigma^2/\{W_1 + B_1 - B_{12}^2/B_2\}$. The within-group estimate of β_1, which corresponds to the restricted randomization estimate has variance σ^2/W_1. Now the difference, $B_1 - B_{12}^2/B_2$, between the denominators in these two variances is the residual sum of squares after fitting a straight line through the means of the x_1 values, with x_2 as 'independent' variable. This means that, if the mean x_1 values plotted against their x_2 values lie close to a straight line, the loss of efficiency of the 'within' estimator is small. If these means lie exactly on a straight line there is no loss of efficiency. This has already been noted in the orthogonal cases treated as in Section 6.2.5.

Example 6.11 In this example the variables are characteristics of timber specimens from $n = 50$ varieties of trees. The data were supplied by the CSIRO Division of Building Research, Melbourne.

$$y \; : \quad \text{modulus of rigidity}$$
$$x_1 : \quad \text{modulus of elasticity}$$
$$x_2 : \quad \text{air dried density}$$

In the following table of the data, the x_2 values are arranged in increasing order of magnitude. Grouped calculations are performed by forming 10 groups each with 5 sets of results and taking all x_2 values within the group to equal the group median.

y	x_1	x_2	y	x_1	x_2
1000	99	25.3	1897	240	50.3
1112	173	28.2	1822	248	51.3
1033	188	28.6	2129	261	51.7
1087	133	29.1	2053	245	52.8
1069	146	30.7	1676	186	53.8
925	91	31.4	1621	188	53.9
1306	188	32.5	1990	252	54.9
1306	194	36.8	1764	222	55.1
1323	195	37.1	1909	244	55.2
1379	177	38.3	2086	274	55.3

y	x_1	x_2	y	x_1	x_2
1332	182	39.0	1916	276	56.9
1254	110	39.6	1889	254	57.3
1587	203	40.1	1870	238	58.3
1145	193	40.3	2036	264	58.6
1438	167	40.3	2570	189	58.7
1281	188	40.6	1474	223	59.5
1595	238	42.3	2116	245	60.8
1129	130	42.4	2054	272	61.3
1492	189	42.5	1994	264	61.5
1605	213	43.0	1746	196	63.2
1647	165	43.0	2604	268	63.3
1539	210	46.7	1767	205	68.1
1706	224	49.0	2649	346	68.9
1728	228	50.2	2159	246	68.9
1703	209	50.3	2078	237.5	70.8

In the model

$$y = \alpha + \beta_1 x_1 + \beta_2 x_2 + \text{error}$$

with independent errors having zero expectations and common variance σ^2, the full least-squares estimate of β_1 is 3.182 with estimated standard error 0.718. Grouping the x_2 values as explained above and calculating a full least-squares estimate gives the result 3.236. Thus the grouping error is quite small relative to the standard error of the estimate.

The within-group least squares estimate of β_1 is 3.046 with an estimated standard error, using the estimated σ^2 from the full least-squares method, of 0.721.

The calculations reported above indicate that both the error due to grouping and the loss of efficiency is small, in this example.

The next step is to find a point estimate of β_1 and a confidence interval, based on the sign statistic S_1. described in Section 6.2.7. A slight, obvious modification of the definition of S_{1s} is required to define S_{1s} in terms of the deviations of the within-group x_1 values from their means. Thus, putting $x_{1.}^s = \sum_{j=1}^{n_s} x_{1s}^s / n_s$,

$$S_{1s}(b) = \sum_{j=1}^{n_s} (x_{1j}^s - x_{1.}^s) \operatorname{sgn} \{ y_{1s,j} - b x_{1j}^s - \hat{D}_s(b) \}$$

where $\hat{D}_s(b) = \text{median } \{y_{1s.j} - bx^s_{1j}, \quad j = 1, 2, \dots, n_s\}$,
and

$$\text{var}\,\{S_{1.}(\beta_1)\} = \sum_{s=1}^{q} C_s \sum_{j=1}^{n_s} (x^s_{1j} - x^s_{1.})^2$$

where $C_s = 1$ or $(n_s - 1)/n_s$ according to whether n_s is even or odd.
In this example every $n_s = 5$, hence

$$\text{var}\,\{S_{1.}(\beta_1)\} = \tfrac{4}{5}W_1 = (212.57)^2$$

The values of $S_{1.}(b_1)$ for a selection of b_1 values are as follows; they give b_1 intervals on which $S_{1.}$ has the values shown.

b_1	$S_{1.}$		
$(\dots, \ 2.10)$	377.5	$(3.83, \ 3.93)$	38.5
$(2.10, \ 2.43)$	328.5	$(3.93, \ 4.05)$	$-\ 58.5$
$(2.43, \ 2.70)$	321.5	$(4.05, \ 4.09)$	$-\ 134.5$
$(2.70, \ 3.23)$	298.5	$(4.09, \ 4.14)$	$-\ 188.5$
$(3.23, \ 3.74)$	241.5	$(4.14, \ 4.31)$	$-\ 224.5$
$(3.74, \ 3.83)$	$-\ 142.5$	$(4.31, \ 4.90)$	$-\ 332.5$
		$(4.90, \ \dots)$	$-\ 432.5$

A plot of $S_{1.}(b_1)$ against b_1 is shown in Fig. 6.4. The abscissae at which the b_1 axis and horizontal lines drawn at $\pm 1.645\,(212.57)$ intersect the graph of $S_{1.}(b_1)$ give the point estimate $\beta_1 = 3.93$ and approximate 90% confidence limits for β_1 of $(2.10, 4.90)$. These results, especially the width of the confidence interval, should be compared with the least-squares results.

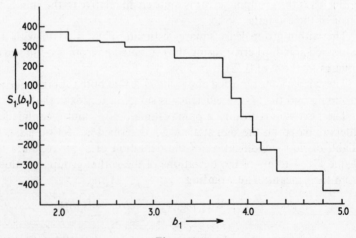

Figure 6.4

6.2.9 *Joint estimation of β_1 and β_2, approximate solutions and estimated standard errors*

Inference about β_1 and β_2 separately has been examined in some detail because it is regarded as an important practical problem. We now return to joint estimation of β_1 and β_2 and consider circumstances where the restricted randomization or grouping methods are not regarded as adequate. For example, the statistician may regard the loss in efficiency as too great, or may be concerned about a nonlinear function of β_1 and β_2, which cannot be handled by the techniques discussed before. We have noted above that while exact joint inference about β_1 and β_2 is possible, inference about β_1 or β_2 alone, or about a function of β_1 and β_2, can be approximate only, in general.

One way of making such inferences is by using standard errors, or large-sample approximate standard errors of estimates. Thus the problem of obtaining estimates of these standard errors arises. There are quite general methods of obtaining estimates of standard errors such as the 'jack-knife' method and the 'bootstrap' method. Here we shall discuss only a method that is directly connected with the actual calculation of the estimates and is related to the well known Newton–Raphson method of solution of equations by successive approximation.

We shall use the rank statistics R_1 and R_2 of Section 6.2.3 in this discussion, but the methods are quite generally applicable.

The basic idea is to approximate the functions $R_i(\mathbf{Y}, \mathbf{b}), i = 1, 2$, by differentiable smooth functions of b_1 and b_2 in the neighbourhood of the solution of the estimating equations (6.11). In order to avoid introducing further notation we shall now treat R_1 and R_2 as differentiable functions of b_1 and b_2. A solution of the estimating equations can now be sought by successsive corrections to a starting solution $\mathbf{b}^\circ$.

Let $C_{st}^\circ = [\partial R_s(b)/\partial b_t]_{\mathbf{b} = \mathbf{b}^\circ}$. Then the first corrections Δb_1° and Δb_2° are obtained by solving the linear equations

$$\begin{bmatrix} C_{11}^\circ & C_{12}^\circ \\ C_{21}^\circ & C_{22}^\circ \end{bmatrix} \begin{bmatrix} \Delta b_1^\circ \\ b_2^\circ \end{bmatrix} = \begin{bmatrix} R_1(\mathbf{Y}, \mathbf{b}^\circ) \\ R_2(\mathbf{Y}, \mathbf{b}^\circ) \end{bmatrix}$$

The new solutions are $b_i^1 = b_i^\circ + \Delta b_i^\circ, i = 1, 2$.

To implement this iterative technique the elements of the matrix $C^\circ = ((C_{st}^\circ))$ are needed. Whereas in the better-known applications of this type of technique the elements of C° are commonly obtained simply by differentiation, here they have to be estimated by a numerical technique. One such technique is to choose $b_i^\circ -$ and $b_i^\circ +$ such that the difference $R_1(\mathbf{Y}, b_1^\circ +, b_2^\circ) - R(\mathbf{Y}, b_1^\circ -, b_2^\circ)$ is approximately equal to the standard deviation of R. Then put

$$C_{11}^\circ \simeq \{R_1(\mathbf{Y}, b_1^\circ +, b_2^\circ) - R_1(\mathbf{Y}, b_1^\circ -, b_2^\circ)\}/(b_1^\circ + - b_1^\circ -)$$

Other C_{st}° are calculated similarly.

The estimated covariance matrix of the estimates $\hat\beta_1, \hat\beta_2$ is given by (6.14) with C_β replaced by $\hat C_\beta$, the estimated covariance matrix, and γ_{rs} replaced by C_{rs}, the latter calculated using $\hat{\boldsymbol\beta}$ instead of $\mathbf{b}^\circ$.

As a possible computational simplification we may note that $\gamma_{12} = 0$ if the design matrix is suitably chosen. In that case the value of C_{rs} may be taken as 0.

Example 6.12 As a numerical illustration we use the data and some of the results in Example 6.11. For preliminary estimates we use those yielded by the sign statistics $S_{1.}$ and $S_{2.}$; they are

$$b_1^\circ = 3.93, \qquad b_2^\circ = 17.76.$$

Also

$$R_1(\mathbf{Y}, 3.93, 17.76) = -3822$$
$$R_1(\mathbf{Y}, 3.83, 17.76) = -2482$$
$$R_2(\mathbf{Y}, 4.05, 17.76) = -5189$$

giving $C_{11}^\circ = -2707/0.21$.

6.2.10 *Joint inference about α and $\boldsymbol\beta$: median regression*

As far as inference about α is concerned, the situation here is exactly as in the straight-line regression case. While $\boldsymbol\beta$ refers only to location shift, the parameter α reflects the actual location. Consequently we are only able to make real progress in a relatively simple manner if we consider median regression, or if the Y variables are assumed to be symmetrically distributed.

Straightforward generalization of the straight-line regression case suggests the use of the following statistics for inference about α and $\boldsymbol\beta$:

$$S_0(\alpha, \boldsymbol\beta) = \sum \text{sgn}(Y_j - \alpha - \beta_1 x_{1j} - \beta_2 x_{2j})$$
$$S_r(\alpha, \boldsymbol\beta) = \sum x_{rj}\, \text{sgn}(Y_j - \alpha - \beta_1 x_{1j} - \beta_2 x_{2j}), \qquad r = 1, 2$$

The joint distribution of S_1, S_2, S_3 has

$$E\{S_r(\alpha, \boldsymbol\beta)\} = 0, \qquad r = 0, 1, 2$$

and the covariance matrix $\mathbf{X'X}$.

With n small it is fairly simple to tabulate the exact joint distribution.

Testing $\alpha = \alpha_0, \beta = \beta_0$

The discussion of Section 5.3.1 is relevant here, and by analogy, a test statistic is

$$Q(\alpha_0, \beta_0) = \mathbf{S}_0' \mathbf{C}_s^{-1}(\alpha_0, \beta_0) \mathbf{S}_0$$

where

$$\mathbf{S}_0' = [S_0(\alpha_0, \beta_0), \quad S_1(\alpha_0, \beta_0), \quad S_2(\alpha_0, \beta_0)]$$

If $\sum x_{1j} = \sum x_{2j} = 0$, which we assume to be true unless otherwise stated, and if also $\sum x_{1j} x_{2j} = 0$, then

$$Q(\alpha_0, \beta_0) = \frac{S_0^2(\alpha_0, \beta_0)}{n} + \frac{S_1^2(\alpha_0, \beta_0)}{\sum x_{1j}^2} + \frac{S_2^2(\alpha_0, \beta_0)}{\sum x_{2j}^2},$$

which is an obvious generalization of expression (5.33).

We have

$$E\{Q(\alpha, \beta)\} = 3$$

and with large n and suitable conditions on the design matrix the distribution of Q may be taken to be approximately χ_3^2. Thus, the observed value of $Q(\alpha_0, \beta_0)$ may be compared with the appropriate χ^2 distribution to test the null hypothesis.

In principle it is, of course, possible to tabulate the null distribution of Q; with n large it is impractical and the χ^2 approximation has to be used.

Example 6.13 Suppose that we have $n = 5$ observations at (x_1, x_2) values $(0, 0), (-1, +1), (+1, -1), (-1, -1), (+1, +1)$

Then the null distribution of Q is as follows:

q :	0.2	1.8	2.2	3.8	4.2	5.0
$16\Pr(Q = q)$:	2	1	4	4	4	1

Point estimation

Point estimates of the parameters are obtained by solving the estimating equations

$$S_r(a, \mathbf{b}) = 0, \qquad r = 0, 1, 2 \tag{6.22}$$

Actually finding solutions can be difficult, but an obvious extension to three dimensions of the type of procedure outlined in Section 6.2.9 can be used to find approximate solutions.

Using the arguments of Section 5.3.1 the large-sample covariance matrix of the estimates obtained as solutions of (6.22) is

$$C_s^{-1}/(4f(0)^2),$$

where $f(0)$ is the density of F at its median.

6.2.11 *Joint inference about α and β: symmetric residuals*

The arguments here are in essence exactly those of Section 5.3.3 in connection with joint inference about α and β in straight-line regression. We condition on the observed absolute residuals $|Y_j - \alpha - \beta_1 x_{1j} - \beta_2 x_{2j}|$ and take these to have positive or negative signs with equal probability. Then putting

$$D_j(\alpha, \boldsymbol{\beta}) = Y_j - \alpha - \beta_1 x_{1j} - \beta_2 x_{2j}$$

we use the statistics

$$A_0(\alpha, \boldsymbol{\beta}) = \sum \operatorname{sgn} D_j(\alpha, \boldsymbol{\beta}) |D_j(\alpha, \boldsymbol{\beta})|$$
$$A_r(\alpha, \boldsymbol{\beta}) = \sum x_{rj} \operatorname{sgn} D_j(\alpha, \boldsymbol{\beta}) |D_j(\alpha, \boldsymbol{\beta})|$$

The exact joint conditional distribution of A_0, A_1, A_2 is in principle easily tabulated but the process becomes impractical as n increases. The joint distribution has

$$E\{A_r(\alpha, \boldsymbol{\beta})\} = 0$$

and covariance matrix

$$\mathbf{C}_A(\alpha, \boldsymbol{\beta}) = \begin{pmatrix} \sum D_j^2 & \sum x_{1j} D_j^2 & \sum x_{2j} D_j^2 \\ \sum x_{1j} D_j^2 & \sum x_{1j}^2 D_j^2 & \sum x_{1j} x_{2j} D_j^2 \\ \sum x_{2j} D_j^2 & \sum x_{1j} x_{2j} D_j^2 & \sum x_{2j}^2 D_j^2 \end{pmatrix}$$

With suitable conditions on the design matrix and the distribution of residuals, this joint distribution is approximately normal for n large.

For joint inference about $\alpha, \boldsymbol{\beta}$, especially testing a hypothesis specifying values of α and $\boldsymbol{\beta}$, we may use the statistic

$$Q_A(\alpha, \boldsymbol{\beta}) = (A_0, A_1, A_2) \mathbf{C}_A^{-1} (A_0, A_1, A_2)'$$

which has expectation exactly 3 and an approximately χ_3^2 distribution under suitable conditions. The point estimates yielded by solving

$$A_r(a, \mathbf{b}) = 0, \qquad r = 0, 1, 2$$

are just the usual least-squares estimates.

Example 6.14 Consider the $n = 5$ observations

x_1	x_2	y	$\lvert y - 1 - x_1 - 2x_2 \rvert$
0	0	1.035	+ 0.035
− 1	+ 1	1.481	+ 0.519
− 1	− 1	− 2.081	− 0.081
+ 1	+ 1	3.611	− 0.389
+ 1	− 1	− 0.938	− 0.938

Testing $H_0 : \alpha = 1, \beta_1 = 1, \beta_2 = 2$

The observed value of Q_A is 2.4962, which is smaller than 3, so that H_0 is accepted.

Enumeration of the 32 possible values of Q shows that $\Pr(Q_A \geqslant 2.4962) = 20/32$.

As we have seen in previous chapters a family of statistics can be generated, starting with A_0, A_1, A_2, by replacing $\lvert D_j \rvert$ by some transformed value $T(\lvert D_j \rvert)$. The simplest of these is $T(u) = 1$ which reproduces the sign statistics S_0, S_1, S_2 that were examined in Section 6.2.10.

Transforming $\lvert D_j \rvert$ to Rank $(\lvert D_j \rvert)$ gives the signed rank statistics that have been studied extensively by Adichie (1967). Writing $R_j =$ Rank $(\lvert D_j \rvert)$, $j = 1, 2, \ldots, n$ it is again relatively straightforward, although tedious for large n, to tabulate the exact joint distribution of the resulting statistics which we denote $A_{Rs}, s = 0, 1, 2$. Although we have transformed to ranks, it will be noted that the conditional joint distribution of $A_{Rs}, s = 0, 1, 2$, is not invariant with the observed sample; this phenomenon was seen in the case of straight-line regression.

Conditionally we have

$$E(A_{Rs}) = 0, \qquad s = 0, 1, 2$$

and the covariance matrix of A_{R0}, A_{R1}, A_{R2} is

$$\mathbf{C}_R = \begin{pmatrix} \sum R_j^2 & \sum x_{1j} R_j^2 & \sum x_{2j} R_j^2 \\ \sum x_{1j} R_j^2 & \sum x_{1j}^2 R_j^2 & \sum x_{1j} x_{2j} R_j^2 \\ \sum x_{2j} R_j^2 & \sum x_{1j} x_{2j} R_j^2 & \sum x_{2j}^2 R_j^2 \end{pmatrix}$$

Unconditionally the covariance matrix is

$$\{ n(n + 1)(2n + 1)/6 \} \mathbf{X}'\mathbf{X}$$

For testing a hypothesis specifying α and β we can use the statistic

$$Q_R(\alpha, \boldsymbol{\beta}) = (A_{R0}, A_{R1}, A_{R2}) \mathbf{C}_R^{-1} (A_{R0}, A_{R1}, A_{R2})'$$

which has properties similar to those of Q_A.

Example 6.15 Using the data of Example 6.14 and testing $H_0 : \alpha = 1$, $\beta_1 = 1, \beta_2 = 2$, the observed value of Q_R is 2.1889, and from the conditional distribution of Q_R we have $\Pr(Q_R \geqslant 2.1889) = 12/16$; therefore H_0 is accepted.

Point estimates based on the signed rank statistics can, in principle, be found as solutions of the estimating equations

$$A_{Rs}(\mathbf{a}, \mathbf{b}) = 0$$

but the comments made in connection with (6.22) regarding finding of solutions also apply here.

6.2.12 *Inference about* α

We shall not give any details about this problem, except to point out that the difficulties discussed in connection with α in straight-line regression arise here as well. Without loss of efficiency it is not, in general, possible to make exact inference about α alone, free of the nuisance parameters β_1 and β_2. Inference about α alone can only be approximate and may have to depend on estimated standard errors, obtained in ways such as those indicated in Sections 5.3.3 and 6.2.9.

6.3 General linear models

If we use the notation of Section 6.2 and put $\gamma = (\alpha, \beta_1, \beta_2)'$ then the model for the location of Y_j can be written

$$\eta = \mathbf{X}'\gamma$$

or equivalently

$$\mathbf{Y} = \mathbf{X}'\gamma + \mathbf{e} \tag{6.23}$$

where $\mathbf{Y} = (Y_1, Y_2, \ldots, Y_n)$ and $\mathbf{e} = (e_1, e_2, \ldots, e_n)$ and $e_1, e_2, \ldots, e_n$ are independent 'errors' distributed identically with distribution function $F(u)$.

The representation of (6.23) is well known for the *general linear model* in which the vector γ is k-dimensional and $\mathbf{X}$ is an $n \times k$ design matrix. In parametric treatments of this model the errors are often taken to be normally distributed. According to our discussion, straight-line and plane regression are obviously special cases of the general linear model. Many statistical techniques that have been developed in other ways are now recognized as special cases of techniques associated with the general linear model. These include, for example, one- and two-way analysis of variance.

The distribution-free methods that have been developed for straight-line and plane regression can be generalized for the general linear model in a fairly obvious way, and we shall discuss some aspects of these generalizations below. However, certain special cases, other than those that we have already considered, are of great practical importance and we shall deal with them separately. They include, for example, one- and two-way analysis of variance.

Rewriting (6.23) as

$$\mathbf{Y} = \mathbf{X}'(\alpha, \boldsymbol{\beta})' + \mathbf{e}$$

where $\boldsymbol{\beta}$ is now a $(k-1)$-dimensional vector, and taking $\mathbf{X}$ to have a first column consisting of ones, we can, in this more general setting, still interpret the parameters $\boldsymbol{\beta}$ as reflecting location shift, whereas α determines the actual locations of the various Y_j distributions. As an illustration we may take the model for the one-way analysis of variance where we have samples of sizes n_i, $i = 1, 2, \ldots, k$, from k populations with location parameters $\mu, \mu + \beta_2, \ldots, \mu + \beta_k$, so that

$$\mathbf{X} = \begin{bmatrix} 1 & 0 & 0 & 0 \\ 1 & 0 & 0 & 0 \\ \vdots & \vdots & \vdots & \vdots \\ 1 & 0 & 0 & 0 \\ 1 & 1 & 0 & 0 \\ \vdots & \vdots & \vdots & \vdots \\ 1 & 1 & 0 & 0 \\ 1 & 0 & 1 & 0 \\ \vdots & \vdots & \vdots & \vdots \\ 1 & 0 & 1 & 0 \\ 1 & 0 & 0 & 1 \\ \vdots & \vdots & \vdots & \vdots \\ 1 & 0 & 0 & 1 \end{bmatrix} \tag{6.24}$$

6.3.1 Inference about $\boldsymbol{\beta}$

By straightforward generalization of statistics used before, the basic set of statistics used in inference about $\boldsymbol{\beta}$ is

$$A_r(\mathbf{Y}, \boldsymbol{\beta}) = \sum_{j=1}^{n} x_{rj}(Y_j - \beta_1 x_{1j} - \ldots - \beta_{k-1} x_{k-1\,j})$$

$$r = 1, 2, \ldots, k-1 \tag{6.25}$$

In this representation we shall assume that, for every $r = 1, 2, \ldots,$ $k - 1, \sum_{j=1}^{n} x_{rj} = 0$. This means that, in the example (6.24) the x-origins are shifted such that in the second column every 0 is replaced by $-n_2/n$ and every 1 by $1 - n_2/n$, and so on.

Under the model according to which the Y_i are identically and independently distributed except for location, all permutations of the realized $D_j(\beta) = Y_j - \beta_1 x_{1j} - \ldots - \beta_{k-1} x_{k-j}, j = 1, 2, \ldots, n$, are equally probable. On this basis it is in theory a simple matter to tabulate the joint conditional distribution of the A statistics. Equally, it is straightforward to tabulate the joint distribution of statistics generated by transformation of the D_j values.

The expectation of every A_r is 0 since we take $\sum x_{rj} = 0$ for every r. The covariance matrix of $A_1, A_2, \ldots, A_{k-1}$ is given by

$$\text{var}(A_r) = \sigma^2 \left(\frac{n}{n-1} \right) \sum_{j=1}^{n} x_{rj}^2$$

$$\text{cov}(A_r, A_s) = \sigma^2 \left(\frac{n}{n-1} \right) \sum_{j=1}^{n} x_{rj} x_{sj}$$

where $\sigma^2 = (1/n) \sum_{j=1}^{n} (d_j - \bar{d})^2, \bar{d} = \left(\frac{1}{n} \right) \sum d_j$. These formulae are obvious generalizations of (6.3).

The inferential problems noted in connection with β_1 and β_2 in the case of plane regression arise here also. Thus it is generally impossible to make exact inferences about individual β's while the others act as nuisance parameters, at least with the permutational scheme being discussed here.

An exact test of a joint hypothesis about $\beta_1, \beta_2, \ldots, \beta_{k-1}$ can be made through the statistic

$$Q_A = \mathbf{A}' \mathbf{C}^{-1} \mathbf{A}$$

where $\mathbf{A}' = (A_1, A_2, \ldots, A_{k-1})$ and $\mathbf{C}$ is the covariance matrix of the A's.

When it is impractical to tabulate the exact distribution of Q_A, we may note that $E(Q_A) = k - 1$ and under suitable conditions on $\mathbf{X}$ and the D_j values the distribution of Q_A will be approximately χ^2_{k-1}.

Transformations of the D_j values in the manner of Section 6.2 produce various statistics that are obvious generalizations of rank and other statistics discussed earlier. For example, let

$$R_r(\mathbf{Y}, \beta) = \sum_{j=1}^{n} x_{rj} \text{Rank} (Y_j - \beta_1 x_{1j} - \ldots - \beta_{k-1} x_{k-1j}),$$
$$r = 1, 2, \ldots, k - 1$$

Then, using the results for A_r we can write down $\text{var}(R_r)$, $\text{cov}(R_r, R_s)$ simply by replacing σ^2 by $(n^2 - 1)/12$.

Example 6.16 The following data relate to trials on percentage elongation of an alloy (Y) and its dependence on the percentages X_1, X_2, X_3, X_4, X_5 of five elements. There were 24 trials ('melts'); the data are from Burr (1974). The y-values are arranged in decreasing order of magnitude and ranks are assigned $24, 23, \ldots, 1$, ties being broken arbitrarily as indicated by the tabulation

y	x_1	x_2	x_3	x_4	x_5	y	x_1	x_2	x_3	x_4	x_5
11.3	0.50	1.3	0.4	3.4	0.010	5.8	0.33	2.6	0.6	4.7	0.008
10.0	0.47	1.2	0.3	3.6	0.012	5.5	0.51	3.9	0.9	4.4	0.000
9.8	0.48	3.1	0.7	4.3	0.000	5.5	0.54	3.1	0.7	4.2	0.000
8.8	0.54	2.6	0.7	4.0	0.022	4.7	0.48	4.0	1.1	3.7	0.024
7.8	0.45	2.8	0.7	4.2	0.000	4.1	0.38	3.3	0.8	4.1	0.000
7.4	0.41	3.2	0.7	4.7	0.000	4.1	0.39	3.2	0.7	4.6	0.016
6.7	0.62	3.0	0.6	4.7	0.026	3.9	0.60	2.9	0.7	4.3	0.025
6.3	0.53	4.1	0.9	4.6	0.035	3.5	0.54	3.2	0.7	4.9	0.022
6.3	0.57	3.7	0.8	4.6	0.000	3.1	0.33	2.9	2.9	1.0	0.063
6.3	0.67	2.7	0.6	4.8	0.013	1.6	0.40	3.2	3.2	1.0	0.059
6.0	0.54	3.1	0.7	4.2	0.000	1.1	0.64	2.5	0.7	3.8	0.018
6.0	0.42	3.1	0.7	4.4	0.000	0.6	0.34	5.0	1.3	3.9	0.044

Standardizing the x variables the covariance matrix of $R_1, \ldots, R_5$ becomes $50\,C_*$, where C_* is the matrix of correlation coefficients among $x_1, x_2, \ldots, x_5$, that is,

$$C_* = \begin{bmatrix} 1.00000 & -0.17600 & -0.42456 & 0.38712 & -0.21924 \\ -0.17600 & 1.00000 & 0.26356 & 0.13389 & 0.20448 \\ -0.42456 & 0.26356 & 1.00000 & -0.87641 & 0.78077 \\ 0.38712 & 0.13389 & -0.87641 & 1.00000 & -0.70099 \\ -0.21924 & 0.20448 & 0.70877 & -0.70099 & 1.00000 \end{bmatrix}$$

$$\text{and } \mathbf{R}' = \begin{bmatrix} 35.1313 \\ -78.7200 \\ -87.3511 \\ 54.1576 \\ -84.2784 \end{bmatrix}$$

Also

$$C_*^{-1} = \begin{bmatrix} 1.3152 & 0.3522 & 0.1962 & -0.6687 & -0.4056 \\ 0.3522 & 2.9125 & -4.4464 & -4.6262 & -0.2896 \\ 0.1962 & 4.4464 & 12.8466 & 10.6459 & -1.6153 \\ -0.6687 & -4.6262 & 10.6459 & 11.6381 & 0.6771 \\ -0.4056 & -0.2896 & -1.6153 & 0.6771 & 2.7061 \end{bmatrix}$$

$$C_*^{-1} \mathbf{R}' = \begin{bmatrix} -0.6906 \\ -54.6376 \\ -52.5560 \\ -13.5852 \\ -41.7494 \end{bmatrix}$$

From these results we obtain

$$Q_R = 9.71$$

which we can treat approximately as an observation on a χ_5^2 random variable on the null hypothesis, H_0, of no association.

Regarding restricted randomization and exact inference about individual β_j values, it is clear that, by this device, exact inference about β_1 will be possible only if we have multiple x_1 values corresponding to at least one of the $(x_2, x_3, \ldots, x_{k-1})$ points.

6.4 One-way analysis of variance

Although the one-way analysis of variance can be treated as a particular case of the general linear model, it is a technique of such practical importance as to deserve special attention. Historically it predates the now well-known general linear approach and its special terminology is useful and very well known.

We consider generalization of the two-sample location problem to one of $k \geqslant 3$ groups of observations from populations that differ only in location. The jth observation from the ith group is denoted y_{ij}, with $j = 1, 2, \ldots, n_i, i = 1, 2, \ldots, k$. The mean of the ith group is $y_{i.}$, the overall mean $y_{..}, \sum_{j=1}^{k} n_j = N$.

The commonly used measure of divergence between the groups is the 'between groups sum of squared differences',

$$B = \sum_{j=1}^{k} n_j (y_{i.} - y_{..})^2$$

and we recall the identity

$$\sum_{k=1}^{k} \sum_{j=1}^{n_i} (y_{1j} - y_{..})^2 = \sum_{i=1}^{k} \sum_{j=1}^{n_i} (y_{ij} - y_{i.})^2 + B$$

in words,

'Total SSD' = 'Within groups SSD' + 'Between groups SSD'

abbreviated

$$T = W + B$$

6.4.1 *The basic randomization test*

Consider the null hypothesis under which the k populations are identical. Then, conditioning on the observed set of y values, the particular partition into k groups of size $n_1, n_2, \ldots, n_k$ can be regarded as one selected at random from all such partitions. Tabulation of the null distribution of B under this permutation scheme is straightforward. Under this scheme T is fixed, hence it is actually immaterial whether we use B or W as a test statistic.

Example 6.17

Group	y_{ij}		
1	5.04,	4.30,	4.90
2	6.49,	5.20	
3	5.38,	3.99	

In this example the observed value of $B = 1.8062$. There are $7!/(3!\,2!\,2!) = 210$ possible partitions, the 10 partitions giving the largest values of B being as follows:

Partitions of the seven observations							B
5.20	5.04	4.90	6.49	5.38	4.30	3.99	3.2042
5.20	5.04	4.90	4.30	3.99	6.49	5.38	3.2042
5.38	5.04	4.90	6.49	5.20	4.30	3.99	2.9114
5.38	5.04	4.90	4.30	3.99	6.49	5.20	2.9114
6.49	5.38	5.20	5.04	4.90	4.30	3.99	2.8793
6.49	5.38	5.20	4.30	3.99	5.04	4.90	2.8793
3.99	4.30	4.90	5.04	5.20	5.38	6.49	2.8564
3.99	4.30	4.90	5.38	6.49	5.04	5.20	2.8564
5.38	5.20	4.90	6.49	5.04	4.39	3.99	2.6964
5.38	5.20	4.90	4.30	3.99	6.49	5.04	2.6964

Therefore $\Pr(\text{observed } B \geqslant 1.8062) > 10/210$ under the null hypothesis H_0 that the three samples are from identical populations. At the 4.8% level we accept H_0.

The listing of possible B values illustrated in Example 6.17 is impractical for moderately large sample sizes and we have to look to approximation of the distribution of B. Under permutation,

$$E(B) = \sum_{i=1}^{k} \frac{n_i}{n_i} \frac{T}{N} \left(\frac{N - n_i}{N - 1} \right) = (k - 1) \left(\frac{T}{N - 1} \right)$$

Putting $B^* = \left(\dfrac{N - 1}{T} \right) B$, we have a statistic whose exact conditional null distribution has expectation $(k - 1)$.

Arguing by analogy to the one-way analysis of variance in the case where the Y_{ij} are normally distributed with variance σ^2, it can be seen that, under reasonable conditions on the Y_{ij}, the permutation distribution of B^* will be approximately χ^2_{k-1}.

We note that B_* is a quadratic form in the random variables $Y_{i.} - Y_{..}$, $i = 1, 2, \ldots, k$, and under permutation

$$\text{var}(Y_{i.} - Y_{..}) = \frac{T}{N n_i} \left(\frac{N - n_i}{N - 1} \right) = \frac{T}{(N - 1)} \frac{1}{n_i} \left(1 - \frac{n_i}{N} \right)$$

$$\text{cov}(Y_{i.} - Y_{..}, Y_{j.} - Y_{..}) = - \frac{T}{(N - 1)} \left(\frac{1}{N} \right)$$

In the usual normal case,

$$\text{var}(Y_{i.} - Y_{..}) = \frac{\sigma^2}{n_i} \left(1 - \frac{n_i}{N} \right)$$

$$\text{cov}(Y_{i.} - Y_{..}, Y_{j.} - Y_{..}) = - \sigma^2 / N$$

hence the covariance structure of the random variables $Y_{i.} - Y_{..}$ is identical under the two schemes. Therefore, if the joint distribution of the $Y_{i.} - Y_{..}$ is approximately normal under permutation, the distribution of B^* is approximately χ^2_{k-1}. The question of approximate normality is the subject of an extension of Theorem 1.8.3.

6.4.2 Rank test (Kruskal–Wallis)

A rank test of the null hypothesis considered in Section 6.4.1 can be made simply by replacing every y_{ij} by its rank. In this special case B^* is denoted B_R^* and we can express it as

$$B_R^* = \frac{12}{N(N + 1)} \sum R_j^2 / n_j - 3(N + 1)$$

where R_j is the sum of the ranks in the jth group.

The conditions for asymptotic χ^2 distribution of B^* apply and with moderately large n_j values it is common to use the χ^2 approximation for the distribution of B_R^*. For selected small values of k and the group sizes it is, of course, possible to tabulate the exact distribution of B^*; see, for example, Lehmann (1975, Table I).

Example 6.18 Ranking the observations in Example 6.17, we find $R_1 = 9, R_2 = 12, R_3 = 7$, giving observed $B_R^* = 2.47$. Enumeration of the partitions of the observations, as in Example 6.17, shows quite quickly that $\Pr(B_R^* \geqslant 2.47) > 0.10$; according to the χ^2 approximation it is close to 0.5.

6.4.3 *Sign test*: $2 \times k$ *contingency test*

Replacing y_{ij} by sgn $(y_{ij} - \hat{y})$, where $\hat{y}$ is the median of all the y results, and writing S_j for the sum of the transformed values in the jth group we note that $T = N$, and $S_1 + S_2 + \ldots + S_k = 0$ if N is even, so that the statistic B^* becomes

$$B_S^* = \left(\frac{N-1}{N}\right) \sum_{j=1}^{k} S_j^2$$

Having transformed the data they can be summarized in a $2 \times k$ contingency table with column totals n_j and both row totals $N/2$. A simple calculation shows that $\{N/(N-1)\} B_S^*$ equals the usual χ_{k-1}^2 test statistic. Minor changes are required if N is odd.

6.5 Two-way analysis of variance

Consider a $b \times t$ array of observations $y_{ij}, i = 1, 2, \ldots, b, j = 1, 2, \ldots, t,$ where typically the results in the same row belong to the same block and the t results within a row are obtained with different treatments. The null hypothesis under test is that there are no differences between treatments.

6.5.1 *The basic randomization test*

Under the null hypothesis, permutations of the realized observations within the rows are equally likely. A conditional test of the null hypothesis can therefore be developed by considering the null distribution of a suitable statistic under the restricted randomization scheme where permutations within rows are allowed.

The treatments sum of squared differences, that is, the between-column sum of squared differences, is commonly used as test statistic. It is defined like B in the one-way analysis of variance; the expression is slightly simpler since each group size is b, thus

$$B = b \sum_{j=1}^{t} (y_{.j} - y_{..})^2$$

where $y_{.j}$ is the mean of the jth column and $y_{..}$ is the overall mean. Let $C = \sum_{i=1}^{b} \sum_{j=1}^{t} (y_{ij} - y_{i.})^2$. Then, under randomization,

$$E(B) = C/b$$

In order to exploit the analogy with the usual two-way analysis of variance we use the statistic

$$B^* = bB(t-1)/C$$

which has expectation $t - 1$ and whose distribution can be approximated by a χ^2_{t-1} distribution under suitable conditions.

6.5.2 Rank test (Friedman)

Transforming the results within each row to their ranks, $1, 2, \ldots, t$, and applying the basic permutation argument to these ranks, the statistic B^* becomes

$$B_R^* = \frac{12}{b(t+1)} \sum_{j=1}^{t} R_j^2 - 3b(t+1)$$

where R_j is the sum of ranks in the jth column.

The exact distribution of B_R^* can be tabulated and tables exist for selected values of b and t; see, for example, Owen (1962). The conditions for applicability of the χ^2_{t-1} approximation hold for B_R^* and it can be used for b and t values not covered by the tables.

6.5.3 Sign test

Let $\hat{y}_i$ be the median of the t observations $y_{i1}, y_{i2}, \ldots, y_{it}$ in the ith row and put $s_{ij} = \mathrm{sgn}(y_{ij} - \hat{y}_i)$, $S_{.j} = S_{1j} + S_{2j} + \ldots + S_{bj}$. Replacing every y_{ij} by its corresponding S_{ij}, the test statistic becomes

$$B_S^* = K \sum_{j=1}^{t} S_{.j}^2$$

where $K = (t-1)/(bt)$ or $1/b$ according as t is even or odd.

The exact distribution of B_S^* can be tabulated for moderately small b and t; for larger b and t the distribution is approximately χ^2_{t-1}. The special case $t = 2$ may be noted as it is the well-known paired two-sample sign test; it is also identical to the rank test when $t = 2$.

Example 6.19 The following table gives a set of results typical of a randomized block experiment. The within-row rank and sign transformed values are also given in each cell, thus: observation, rank, sign.

		\multicolumn{4}{c}{Treatments ($t = 4$)}
		T_1	T_2	T_3	T_4
Blocks	B_1	2.2, 3, $+1$	3.4, 4, $+1$	1.7, 1, -1	1.9, 2, -1
($b = 3$)	B_2	4.6, 4, $+1$	3.7, 3, $+1$	2.8, 2, -1	2.7, 1, -1
	B_3	1.5, 1, -1	2.7, 4, $+1$	1.9, 2, -1	2.0, 3, $+1$

The values of the three statistics discussed in Sections 6.5.1, 6.5.2, 6.5.3 are as follows

$$B^* = 4.72$$
$$B_R^* = 4.20$$
$$B_S^* = 5.00$$

It should be noted that if tied ranks occur the computations for B_R^* can be done according to the method for B^*. Although formulae for B_R^* can be written down they are not really necessary.

6.5.4 *Aligned ranks and other scoring systems*

The transformations indicated in Sections 6.5.2 and 6.5.3 may be regarded as two systems of converting the original observations to 'scores' before performing the randomization test. Many others have been used, an interesting class of scoring systems being that in which a transformation is not made within rows separately, but overall, after eliminating block effects. For example, we may first calculate row (block) averages and then subtract them from the values within the block. This process is also called 'aligning'. After aligning, the entire set of adjusted observations is ranked $1, 2, 3, \ldots, N$, with $N = bt$. The analysis then proceeds as before, but note that we still consider only within-row permutations.

As indicated in Lehmann (1975, p. 270), there may be some advantage in ranking after aligning; this is very clear for $t = 2$. The statistic B^* is unchanged by aligning according to row means, thus making any real gains with moderately large values of t problematical. There is also a question of the method to be used for aligning. For example, the median could be used instead of the mean. Finally, if the results within a row are affected by random errors, apart from random effects due to the allocation of treatments to plots, aligning does not necessarily lead to exact test procedures.

EXERCISES

6.1 The data in the table below are observations taken at the value (x_1, x_2) of a 'dependent variable' according to the plane regression model of Section 6.2.

	x_2				
	-2	-1	0	1	2
-2	3.5	2.8	8.8	10.8	11.6
-1	6.7	7.2	8.3	10.4	13.1
x_1 0	5.7	7.0	10.4	11.0	13.6
1	5.5	6.9	10.1	11.1	17.6
2	8.2	8.3	13.9	12.0	14.8

Using a rank method test the null hypothesis $H_0 : \beta_1 = 1$, $\beta_2 = 2$ against H_1 : not H_0 at the 10% level of significance.

6.2 Using the data in Exercise 6.1 obtain exact $100\gamma\%$ confidence intervals for each of β_1 and β_2 with $\gamma \simeq 0.9$ based on sign statistics.

6.3 Consider an (x_1, x_2) array as in Exercise 6.1 with both x_1 and x_2 having values $-k, -k+1, \ldots, -1, 0, 1, 2, \ldots, k-1, k$, and suppose that the distribution of the Y-residuals, F, is a mixture of two normal distributions $N(0, 1), N(0, 4)$ in proportions 0.8 and 0.2.

Obtain large k values of the variances of the estimates of β_1 and β_2 based on the sign statistics used in Exercise 6.2. Compare these variances with those of the least squares estimates of β_1 and β_2.

6.4 In the one-way analysis of variance with $k = 3$ groups let

$$A_i = \sum_{j=1}^{n_i} y_{ij} - n_i y_{..}, \quad i = 1, 2, 3.$$

Using results given in Section 6.4, or otherwise, obtain the covariance matrix under permutation of A_1 and A_2, and denote it by V.

Show that $Q = (A_1, A_2)V^{-1}(A_1, A_2)$ is identical to B^*.

6.5 In Exercise 6.4, let $R_i = \sum_{j=1}^{n_i} \text{Rank}(y_{ij})$ as in the definition of the Kruskal–Wallis test, and let Q_R denote the statistic analogous to Q based on R_1 and R_2. Write

$$A_i(t_1, t_2) = \sum_{j=1}^{n_i} (y_{ij} - t_i) - n_i y_{..}(t_1, t_2), i = 1, 2$$

where $y_{..}(t_1, t_2)$ denotes the mean of the 'adjusted' values $(y_{11} - t_1) \ldots, (y_{21} - t_2), \ldots y_{31}, \ldots y_{3n_3}$, and define $R_i(t_1, t_2)$ similarly.

Obtain expressions for the efficacies of the statistics Q and Q_R using the definition in Section 1.7.3, Chapter 1, and find the ratio of the squared efficacies. Compare the result with the analogous result for the two sample case.

6.6 Parabola regression of the form '$y = \alpha + \beta_1 x + \beta_2 x^2$' can be considered as a special case of plane regression with $x_{1i} = x_i$, $x_{2i} = x_i^2$, where x_{1i} and x_{2i} are the variables used in Section 6.2.

The data below can be treated as relating to an experiment on a quantitative factor whose level is x, the three values of x used in the experiment being 0, 1, 2. At each value of x four independent y-observations are made.

| | x | |
0	1	2
8.13	13.81	21.80
9.49	12.65	20.20
9.78	13.20	19.98
13.28	12.49	19.58

Calculate the value of the appropriate quadratic form based on ranks for testing $H_0: \beta_1 = 1, \beta_2 = 2$, and state your conclusion.

6.7 Use the data in Exercise 6.6, assume $\beta_1 = 1$ and obtain a confidence interval for β_2 with confidence coefficient approximately 0.95.

6.8 The following data from Burr (1974) p. 369 are measurements of diameter of pieces of 4 inch stock machined on a lathe. Measurements were made at five positions on each piece, and are recorded as (diameter $- 3.96) \times 10^4$.

	Position				
Piece	1	2	3	4	5
1	72	71	73	70	68
4	76	72	70	71	70
7	76	73	72	72	72
10	75	75	73	73	73

Interpreting 'pieces' as blocks and 'positions' as treatments, perform a Friedman rank test of significance of position.

6.9 Assuming the positions in Exercise 6.8 to be equally spaced a plot of the data suggests that there may be a linear trend within each block. Using the method of Section 6.2.4, obtain a point estimate of the slope parameter of this trend, assuming it to be common to the four blocks.

CHAPTER 7

Bivariate problems

7.1 Introduction

Two types of question arising naturally in connection with bivariate distributions are those of association and of location; although problems of dispersion were discussed briefly in the univariate case we shall ignore them in this chapter. In certain fields of application, like psychology, tests and measures of association are often used, and the first part of this chapter will be devoted to some of the well-known relevant methods. In the later part we shall look at some questions of location. Multivariate statistics are more difficult to interpret than univariate statistics, and so we restrict attention to the bivariate case, although similar methods can be applied to higher-dimensional random variables.

7.2 Tests of correlation

7.2.1 *Conditional permutation tests: the product moment correlation coefficient*

We have a sample (x_i, y_i), $i = 1, 2, \ldots, n$, of n independent observations on the bivariate random variable (X, Y). In many applications the question of possible independence of X and Y is considered and statistics related to the correlation coefficient are often used in this connection. Technically a test of zero correlation is not, of course, a test of independence, but where such a test is used the primary concern is often to establish dependence, rather than independence.

The basis of the tests that we shall discuss is that if X and Y and independent and the observed x_i values are arranged in order of magnitude, then the corresponding sequence of y_i values should behave like a random sequence. Consequently the tests of randomness mentioned in connection with inference about the slope in straight-line regression are applicable here, if we condition on the set of observed x_i values.

The basic statistic considered in straight-line regression, with slope $\beta = 0$ is of the form $T = \sum x_i y_i$. Keeping $x_1, x_2, \ldots, x_n$ fixed and considering permutations of the y_i values we have seen, in Chapter 5, that the permutation distribution of T can be tabulated quite easily for small n and that

$$E(T) = n\bar{x}\bar{y}$$

$$\text{var}(T) = \{\sum(y_i - \bar{y})^2\}\{\sum(x_i - \bar{x})^2\}/(n - 1)$$

with the distribution approaching normality as n increases, under suitable conditions on the x_i and y_i sequences.

In the present context the sample correlation coefficient R, which is a linear function of T is often used; in our present notation,

$$R = \frac{T - n\bar{x}\bar{y}}{\{\text{var}(T)\}^{1/2}\sqrt{(n - 1)}}$$

so that we have for the permutation distribution of R,

$$E(R) = 0, \quad \text{var}(R) = 1/(n - 1)$$

The expression for R is symmetric in the x_i and y_i; if the conditioning used above is on fixed y_i instead of x_i values the result is exactly as before.

7.2.2 Spearman rank correlation

If we transform the x_i values and the y_i values to their respective ranks and then calculate R as above, we obtain the Spearman rank correlation coefficient. Its distribution under the hypothesis of independence of X and Y can be tabulated according to the same principle used in Section 7.2.1. Owing to the transformation to ranks, this null distribution is the same for every sample of the same size, its tabulation is relatively easy, and can be done once and for all; tables of the null distribution of the Spearman rank correlation coefficient can be found in many books on non-parametric statistics.

Example 7.1

x_i :	7.6	2.3	6.5	Rank (x_i) :	3	1	2
y_i :	6.3	3.8	6.6	Rank (y_i) :	2	1	3

$$R_S = \frac{13 - (6 \times 6)/3}{\{(14 - 12)(14 - 12)\}^{1/2} \sqrt{2}} = \frac{1}{2}$$

In this example the null distribution of R_S can be obtained by listing the 6 permutations of the y_i ranks against the fixed ranks 1, 2, 3 of x_i values, and we find

$$\Pr(R_S = 1) = \Pr(R_S = -1) = 1/6, \Pr(R_S = \tfrac{1}{2}) = \Pr(R_S = -\tfrac{1}{2}) = 2/6.$$

After the rank transformations the numerator of R_S is

$$\sum \text{Rank}(x_i) \cdot \text{Rank}(y_i) - n(n + 1)^2/4$$

while the denominator is $n(n^2 - 1)/12$. Putting $D_i = \text{Rank}(x_i) - \text{Rank}(y_i)$, we see that

$$\sum D_i^2 = n(n + 1)(2n + 1)/3 - 2\sum \text{Rank}(x_i) \text{Rank}(y_i)$$

and

$$R_S = 1 - 6\sum D_i^2/\{n(n^2 - 1)\},$$

an often-used expression.

From Section 7.2.1 we have immediately that, in the null case,

$$E(R_S) = 0, \quad \text{var}(R_S) = 1/(n - 1)$$

and a normal approximation can be used for the distribution of R_S for large n.

When the hypothesis of independence of X and Y is rejected, the value of R_S is often used as a measure of the strength of the association between X and Y. Here we do not have the useful interpretation of the ordinary correlation coefficient, for homoscedastic regression, relating the marginal and conditional variables of Y. However, if the plot of y ranks against x ranks reveals a linear trend, we do have approximately

$$\text{Mean}\{\text{Rank}(Y)|\text{Rank}(X) = x\} = R_S\{x - (n + 1)/2\} + (n + 1)/2$$
$$\text{s.d.}\{\text{Rank}(Y)|\text{Rank}(X) = x\} = \{(1 - R_S^2)(n^2 - 1)/12\}^{1/2}$$

which may give a fairly useful prediction of a Y-rank from an observed X-rank.

7.2.3 Sign transformations: 2×2 contingency tables

Instead of transforming x_i and y_i values to their ranks before calculating the correlation coefficient, we may make the sign transformations $\text{sgn}(x_i - \hat{x})$, $\text{sgn}(y_i - \hat{y})$, where $\hat{x}$ and $\hat{y}$ are the respective sample medians. The results can then be summarized in a 2×2 table as follows, if n is even.

$$\text{sgn}(X - \hat{x})$$

		-1	$+1$	
$\text{sgn}(Y - \hat{y})$	-1	a	b	$n/2$
	$+1$	c	d	$n/2$
		$n/2$	$n/2$	n

The correlation coefficient can then be expressed as

$$R_C = (a + d - b - c)/n$$

and simple calculations confirm that $R_C \sqrt{(n-1)}$ is the normal deviate that arises in the exact (Fisher) test for association in a 2×2 contingency table when a normal approximation is used for the relevant hypergeometric distribution.

7.2.4 *Kendall's τ and Theil's statistic*

Suppose that we have arranged the x values in increasing order of magnitude so that $x_1 < x_2 \ldots < x_n$. Then Theil's statistic for testing randomness of the y sequence is

$$T = \sum_{i < j} \text{sgn}(y_j - y_i)$$

(see also Section 5.2.6). The Kendall rank correlation coefficient is

$$\tau = 2T/\{n(n-1)\}$$

Details of the tabulation of the null distribution of T are given in in Chapter 5, where it is also shown that it has $E(T) = 0$, $\text{var}(T) = n(n-1)(2n+5)/18$, and is asymptotically normally distributed. Thus

$$E(\tau) = 0, \quad \text{var}(\tau) = 8(2n+5)/\{9n(n-1)\}$$

An interesting interpretation of Kendall's τ statistic is that the proportion of (x_i, y_i), (x_j, y_j) pairs in which the y order is the same as the x order is $(1 + \tau)/2$.

7.2.5 *Mean square successive difference test*

Correlation tests are known to be insensitive to certain departures from independence; hence it is worth while considering some other tests. One of these is an adaptation of the test of trend based on the mean square successive difference. The idea of the test is that dependence between X and Y may be indicated by the conditional variance of Y given X being, on average, substantially smaller than the marginal variance of Y. Therefore, if the observed X values are fixed in increasing order the mean square successive difference of the Y sequence would be relatively low, if the dependence takes the form that the conditional expectation of Y given X follows a smooth trend in X.

Tabulation of the conditional distribution of the statistic

$$M = \sum_{i=1}^{n-1} (y_{i+1} - y_i)^2$$

where we take $x_1 < x_2 \ldots < x_n$, is straightforward and requires only listing of all permutations of the y values. The first and second moments of M can be expressed in terms of

$$\mu_r = \sum_{i=1}^{n} (y_i - \bar{y})^r / n, \quad r = 2, 4$$

where $\bar{y} = \sum_{i=1}^{n} y_i / n$
We have

$$E(M) = 2n\mu_2$$

while for n large the expression for $E(M^2)$ simplifies to

$$E(M^2) \simeq 3n\mu_4 + 7n^2\mu_2^2$$

and these quantities can be used in an approximate test of independence of X and Y.

One of the less attractive features of the test based on M is that the statistic is not symmetric in the x and y observations.

7.2.6 Contingency tables, correlation ratios

When n items are cross-classified in an $r \times c$ array according to two criteria A and B, the table of frequencies is usually called an $r \times c$ contingency table. The usual χ^2 test of significance associated with contingency tables can be interpreted as a test of independence of the classifying mechanisms A and B.

If the classifications A and B are qualitative the type of analysis that usually accompanies contingency tables seems quite satisfactory. However, if A and B actually correspond to random variables X and

Y for which some form of grouping has been chosen so that the contingency table is a bivariate frequency distribution of X and Y, then the contingency table analysis seems somewhat less satisfactory, since it depends on an arbitrary grouping, and takes no account of the orderings of A and B.

The correlation ratios (Kendall and Stuart, 1961, VII, p. 296) which are developed from a partitioning of the Y-sum of squared differences, in the manner of the analysis of variance, can also be used to test independence. For data on continuous X and Y grouped into a bivariate frequency table, the null distributions of these coefficients, under permutation, can be obtained by conditioning on the appropriate marginal totals. These coefficients are not symmetric in the x and y values and depend on the grouping employed for the bivariate frequency tables.

7.3 One-sample location

7.3.1 *Medians*

We now suppose that a random sample (x_i, y_i), $i = 1, 2, \ldots, n$, has been drawn from the continuous bivariate distribution of (X, Y). We shall denote the median of X by θ_x and that of Y by θ_y. Further, let $\Pr(X > \theta_x, Y > \theta_y) = \pi_{11}$. Then

$\Pr(X \leqslant \theta_x, Y > \theta_y) = 1/2 - \pi_{11} = \Pr(X > \theta_x, Y \leqslant \theta_y)$, and $\Pr(X \leqslant \theta_x, Y \leqslant \theta_y) = \pi_{11}$. The symmetries
$\Pr(X > \theta_x, Y > \theta_y) = \Pr(X \leqslant \theta_x, Y \leqslant \theta_y)$ and
$\Pr(X > \theta_x, Y \leqslant \theta_y) = \Pr(X \leqslant \theta_x, Y > \theta_y)$, conversely, imply that
θ_x and θ_y are the respective medians.

Testing a specified (θ_x, θ_y)

The symmetry noted above can be exploited to develop an exact test of a hypothesis specifying two values $(\theta_x^\circ, \theta_y^\circ)$. Suppose that in the observed sample,

$$n_{00} = \#\,(x_i \leqslant \theta_x^\circ, y_i \leqslant \theta_y^\circ), n_{01} = \#\,(x_i \leqslant \theta_x^\circ, y_i > \theta_y^\circ),$$
$$n_{10} = \#\,(x_i > \theta_x^\circ, y \leqslant \theta_y^\circ), n_{11} = \#\,(x_i > \theta_x^\circ, y_i > \theta_y^\circ)$$

If we now condition on both $n_{00} + n_{11}$, and $n_{10} + n_{01}$ being fixed at the observed values, then n_{00} and n_{10} are, conditionally, distributed $\mathrm{Bin}(n_{00} + n_{11}, 1/2)$ and $\mathrm{Bin}(n_{10} + n_{01}, 1/2)$, and they are independent. As a test statistic we may use

$$Q = \frac{4[n_{00} - (n_{00} + n_{11})/2]^2}{(n_{00} + n_{11})} + \frac{4[n_{10} - (n_{10} + n_{01})/2]^2}{(n_{10} + n_{01})}$$

(see also Section 1.7.3 for similar statistics in general). The expression for Q can be written more simply as $Q = (n_{00} - n_{11})^2/(n_{00} + n_{11}) + (n_{10} - n_{01})^2/(n_{10} + n_{01})$, but in the first form it is easier to see that the appropriate continuity correction when using normal and χ^2 approximations is to replace $|n_{00} - (n_{00} + n_{11})/2|$ by $|n_{00} - (n_{00} + n_{11})/2| - 1/2$, the other numerator being treated similarly.

Since we know that the binomial $(n, \frac{1}{2})$ distribution can be approximated by a normal distribution for moderately large n, we shall be able to approximate the distribution of Q by a χ_2^2 distribution. Tabulation of the exact distribution of Q is straightforward but since the normal approximation to $\text{Bin}(n, \frac{1}{2})$ is good for quite small values of n, the approximation by χ_2^2 should be adequate for most applications.

Example 7.2 Suppose that a sample of $n = 200$ (x, y) pairs with given values of $\theta_x^\circ, \theta_y^\circ$ yields

$$n_{00} = 71, n_{11} = 69, n_{01} = 40, n_{10} = 20$$

Then applying continuity corrections,

$$Q = (|40 - 20| - 1)^2/60 + (|71 - 69| - 1)^2/140 = 6.03$$

Referring this Q to χ_2^2, we reject the hypothesized $(\theta_x^\circ, \theta_y^\circ)$ at the 2.5% level.

If the two median values were tested individually, both would be accepted at the 10% level. The frequencies given above indicate that there is strong dependence between X and Y, and the results of the significance tests illustrate the well-known fact that multivariate tests are more sensitive than univariate tests under suitable conditions.

Confidence regions for (θ_x, θ_y)

A joint confidence region for (θ_x, θ_y) can be set by using Q and taking a trial point (θ_x, θ_y) to belong to the confidence region if the corresponding value of Q is less than an appropriately chosen constant. The following example illustrates the procedure.

Example 7.3 The $n = 30$ pairs of (x, y) observations listed below were generated from a bivariate normal distribution with correlation coefficient 0.8.

x	y	x	y	x_1	y_1
0.448	− 0.435	− 0.688	− 0.668	− 0.173	− 0.622
− 0.062	− 0.831	− 0.199	− 1.029	− 0.644	− 1.282
0.241	0.313	− 0.428	0.093	− 0.587	− 0.372
− 2.181	− 2.495	− 2.318	− 2.707	− 1.541	− 0.201
− 0.456	− 0.500	− 0.077	− 0.108	− 1.256	0.169
1.262	1.395	− 0.987	− 1.702	0.542	1.012
0.385	− 0.215	− 0.110	− 0.227	− 0.788	− 0.978
0.207	0.585	− 2.523	− 2.515	1.219	0.482
− 1.185	− 1.028	0.514	0.773	− 0.276	− 0.819
− 1.320	− 0.746	− 0.363	− 1.014	− 0.213	0.103

The data points are shown in Fig. 7.1. We illustrate determination of a confidence region by considering two (θ_x, θ_y) trial values

$$\theta_x = -0.2, \theta_y = -0.6 : n_{00} = 11, n_{01} = 7, n_{11} = 9, n_{10} = 3$$
$$Q = 0.95$$

$$\theta_x = -0.4, \theta_y = -0.9 : n_{00} = 7, n_{01} = 7, n_{11} = 14, n_{10} = 2$$
$$Q = 3.49$$

Both Q values are calculated with a correction for continuity. With $\chi_2^2(0.80) = 3.22$, the first of the two points belongs to the 80% confidence region and the second does not. Proceeding with similar calculations, it is an easy task to obtain an outline of the joint 80% confidence region for (θ_x, θ_y); the shaded area in Fig. 7.1 shows such a region in approximate outline. To construct a confidence region note that as θ_x and θ_y vary, the value of Q changes only when θ_x passes through one of the observed x values, and when θ_y passes through one of the observed y values.

The ellipse sketched in Fig. 7.1 is an 80% confidence region for (θ_x, θ_y) based on normal theory; the reason for choosing confidence coefficient 80% is that tangents to the ellipse drawn parallel to the axis give two-sided 93% confidence limits (by interpolation in F- and t-tables) for the individual θ_x and θ_y; this confidence coefficient is close to those used conventionally. In the case of the distribution-free limits, tangents drawn to the confidence contour also give confidence limits with coefficient greater than 80%, but the exact relation is not known.

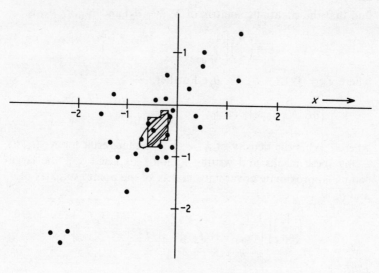

Figure 7.1

However, it will be seen that the tangents to the 80% distribution-free confidence region are close to the individual two-sided 93% confidence limits.

Point estimation

Let $N_{00}(t_x, t_y) = \#(X_i \leqslant t_x, Y_i \leqslant t_y)$, etc., in a notation corresponding to that used above. Since $E\{N_{00}(\theta_x, \theta_y) - N_{11}(\theta_x \theta_y)\} = 0$ and $E\{N_{10}(\theta_x, \theta_y) - N_{01}(\theta_x \theta_y)\} = 0$, we estimate θ_x and θ_y by solving the equations in t_x and t_y:

$$N_{00}(t_x, t_y) - N_{11}(t_x, t_y) = 0$$
$$N_{10}(t_x, t_y) - N_{01}(t_x, t_y) = 0$$

As we have indicated in the introduction to this section, these equations are equivalent to

$$N_{00}(t_1, t_2) + N_{01}(t_1, t_2) = N_{0.}(t_x, t_y) = n/2$$
$$N_{00}(t_1, t_2) + N_{10}(t_1, t_2) = N_{.0}(t_x, t_y) = n/2$$

Since $N_{0.}(t_x, t_y)$ does not depend on t_y, and $N_{.0}(t_x, t_y)$ not on t_x, the relevant solutions are just the two-sample medians.

Using well-known results about the multinomial distribution, we find that the covariance matrix of $N_{0.}(\theta_1, \theta_2)$ and $N_{.0}(\theta_1, \theta_2)$ is

$$V = \left(\frac{n}{4}\right)\begin{bmatrix} 1 & 4\pi_{00} - 1 \\ 4\pi_{00} - 1 & 1 \end{bmatrix}$$

where $\pi_{00} = \Pr(X < \theta_x, Y < \theta_y)$. Further,

$$[\partial E\{N_{0.}(t_x, t_y)\}/\partial t_w]_{t_x = \theta_x, t_y = \theta_y} = \begin{cases} n f_x(\theta_x), & w = x \\ 0 & w = y \end{cases}$$

where $f_x(x)$ is the density of X, with a similar result for $N_{.0}(t_x, t_y)$. Using these results and writing f_w for $f_w(\theta_w)$, $w = x, y$, the large-sample approximate covariance matrix of the point estimates of θ_x and θ_y is

$$\left(\frac{1}{4n}\right)\begin{bmatrix} 1/f_x^2 & (4\pi_{00} - 1)/f_x f_y \\ (4\pi_{00} - 1)/f_x f_y & 1/f_y^2 \end{bmatrix}$$

Efficacy

According to the definition of efficacy in the two-parameter case, we have for the bivariate median test

$$\begin{aligned}(\text{Efficacy})^2 &= n(f_x f_y) V^{-1}(f_x, f_y)' \\ &= 4\{f_x^2 + f_y^2 + 2f_x f_y(1 - 4\pi_{00})\}/\{1 - (4\pi_{00} - 1)^2\}\end{aligned}$$

Example 7.4 Suppose that the joint distributions of X and Y is normal with $E(X) = E(Y) = 0$, $\text{var}(X) = \text{var}(Y) = 1$ and correlation coefficient ρ. Then writing $p_m = 4\pi_{00} - 1$ we have $p_m = (2/\pi)\sin(\rho)$ and the efficacy of the median test is expressed as

$$\{\text{efficacy (median)}\}^2 = (2/\pi)/\{2/(1 + p_m)\}$$

For the test based on the sample means

$$\{\text{efficacy (mean)}\}^2 = 2/(1 + \rho)$$

giving

$$\{\text{efficacy (median)/efficacy (mean)}\}^2 = (2/\pi)(1 + \rho)/(1 + p_m)$$

(see also Puri and Sen (1971), p. 175).

7.3.2 *Symmetric distributions*

Denoting the joint density of X and Y by $F(x, y)$, we now take the distribution of (X, Y) to be symmetric about (θ_x, θ_y) in the sense that

$$F(\theta_x + \xi, \theta_y + \eta) = F(\theta_x - \xi, \theta_y - \eta)$$

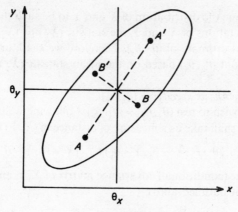

Figure 7.2

Referring to Fig. 7.2, this means that at points A and A' the densities are identical, and similarly for B and B'. It will be noted that the bivariate normal distribution is symmetric in this sense.

With the symmetry that we have assumed, the basic permutation argument that we shall use to make inferences about (θ_x, θ_y) rests on the following conditioning: Let the jth observational pair be (x_j, y_j) and put $\xi_j = x_j - \theta_x, \eta_j = y_j - \theta_y$. We now consider all sets of possible observations such that the jth observational pair is

$$(\theta_x + \xi_j, \theta_y + \eta_j) \quad \text{or} \quad (\theta_x - \xi_j, \theta_y - \eta_j)$$

with equal probability. Fixing attention on the x values only, the conditioning is seen to be the same as that used in the univariate case, namely on the $|x_j - \theta_x|$ values being fixed; however, in the bivariate case the observed signs of ξ_j and η_j have to be noted.

To simplify notation, suppose for the moment that $(\theta_x, \theta_y) = (0,0)$. Adjustment for $(\theta_x, \theta_y) \neq (0,0)$ is straightforward. We shall be using the conditional distributions of statistics generated by taking the 2^n sets of observations

$$\pm(x_1, y_1), \pm(x_2, y_2), \ldots, \pm(x_n, y_n)$$

as having equal probabilities. Conditionally, therefore, we shall be dealing with n independent random variables (X_i, Y_i), where

$$\Pr(X_i = x_i, Y_i = y_i) = \Pr(X_i = -x_i, Y_i = -y_i) = 1/2, \quad i = 1, 2, \ldots, n$$

Among the statistics that we shall consider is $(X_\cdot, Y_\cdot) = (\sum X_i, \sum Y_i)$. Under the permutation scheme $E(X_\cdot) = E(Y_\cdot) = 0$, and

$$\text{var}(X_\cdot) = \sum x_i^2, \quad \text{var}(Y_\cdot) = \sum y_i^2$$

with $\quad \text{cov}(X_\cdot, Y_\cdot) = \sum x_i y_i$

If we assume the distribution of X and Y to be such that θ_x and θ_y are the expectations of X and Y, the statistic $(\bar{X}, \bar{Y}) = (X./n, Y./n)$ is a natural point estimate of (θ_x, θ_y). Therefore we shall, first, consider inference about (θ_x, θ_y) based on the mean statistic $(\bar{X}, \bar{Y})$.

Testing a hypothesis specifying (θ_x, θ_y)
Suppose we wish to test $(\theta_x, \theta_y) = (\theta_x^\circ, \theta_y^\circ)$. Following well-established tradition we shall take as a measure of distance of $(\bar{X}, \bar{Y})$ from $(\theta_x^\circ, \theta_y^\circ)$

$$Q(\theta_x^\circ, \theta_y^\circ) = (\bar{X} - \theta_x^\circ, \bar{Y} - \theta_y^\circ) V^{-1} (\bar{X} - \theta_x^\circ, \bar{Y} - \theta_y^\circ)' \qquad (7.1)$$

where V is the (conditional) covariance matrix of $\bar{X}, \bar{Y}$, and following the argument outlined above,

$$n \, \mathrm{var}(\bar{X}) = \sum (x_i - \theta_x^\circ)^2, \quad n \, \mathrm{var}(\bar{Y}) = \sum (y_i - \theta_y^\circ)^2,$$
$$n \, \mathrm{cov}(\bar{X}, \bar{Y}) = \sum (x_i - \theta_x^\circ)(y_i - \theta_y^\circ) \qquad (7.2)$$

In principle we have to list all 2^n possible sets and calculate the corresponding values of Q for the observed (x_i, y_i) values.

Example 7.5 Suppose we have the $n = 3$ pairs

$$(-1.0, +0.4), (-0.1, +1.5), (+0.2, +0.3)$$

and we wish to test $H_0 : (\theta_x, \theta_y) = (0.5, 1.0)$.

The observed $(x_i - \theta_x^\circ, y_i - \theta_y^\circ)$ values are

$$(-1.5, -0.6), (-0.6, +0.5), (-0.3, -0.7)$$

giving

$$nV = \begin{pmatrix} 2.70 & 0.81 \\ 0.81 & 1.10 \end{pmatrix}$$

and the observed $Q(0.5, 1.0) = 2.1408$.

A list of the $2^3 = 8$ possible sets of observations and their corresponding Q values follows:

Sample values			Q
$(-1.5, \ -0.6)$	$(-0.6, \ +0.5)$	$(-0.3, \ -0.7)$	2.1408
$(+1.5, \ +0.6)$	$(-0.6, \ +0.5)$	$(-0.3, \ -0.7)$	0.1898
$(-1.5, \ -0.6)$	$(+0.6, \ -0.5)$	$(-0.3, \ -0.7)$	2.9529
$(+1.5, \ +0.6)$	$(+0.6, \ -0.5)$	$(-0.3, \ -0.7)$	2.7164
$(-1.5, \ -0.6)$	$(-0.6, \ +0.5)$	$(+0.3, \ +0.7)$	2.7164
$(+1.5, \ +0.6)$	$(-0.6, \ +0.5)$	$(+0.3, \ +0.7)$	2.9529
$(-1.5, \ -0.6)$	$(+0.6, \ -0.5)$	$(+0.3, \ +0.7)$	0.1898
$(+1.5, \ +0.6)$	$(+0.6, \ -0.5)$	$(+0.3, \ +0.7)$	2.1408

Thus $\Pr(\text{observed } Q \geqslant 2.1408) = 6/8$, and we accept H_0. In fact, with this sample size the smallest significance level at which we could reject a null hypothesis is $2/8$.

The permutation distribution of $Q(\theta_x^{\circ}, \theta_y^{\circ})$ has mean exactly 2. Under suitable conditions on the joint distribution of X and Y the distribution of Q may be approximated by a χ_2^2 distribution for n sufficiently large.

Confidence regions

As usual we may determine a confidence region as the set of all possible (θ_x, θ_y) points which are acceptable according to a hypothesis test at a fixed level. In an exact procedure this would entail evaluation of the exact null distribution of Q for every (θ_x, θ_y).

If we approximate the distribution of Q by a χ_2^2 distribution an approximately $100\beta\%$ confidence region is obtained as all (θ_x, θ_y) satisfying

$$\frac{\begin{aligned}(\bar{x} - \theta_x)^2 \sum (y_i - \theta_y)^2 + (\bar{y} - \theta_y)^2 \sum (x_i - \theta_x)^2 - 2(\bar{x} - \theta_x) \\ \times (\bar{y} - \theta_y) \sum (x_i - \theta_x)(y_i - \theta_y)\end{aligned}}{[\{\sum (x_i - \theta_x)^2\}\{\sum (y_i - \theta_y)^2\} - \{\sum (x_i - \theta_x)(y_i - \theta_y)\}^2]} \leqslant \chi_2^2(\beta)$$

If we use the equality in the expression above and fix θ_y, we produce a quadratic equation in θ_x, so we can compute the confidence region fairly easily.

7.3.3 *Symmetric distributions: transformation of observations*

Employing a method of writing that we have first used in the univariate case, we put

$$(x_i - \theta_x, y_i - \theta_y) = [\text{sgn}\,(x_i - \theta_x)|x_i - \theta_x|, \text{sgn}\,(y_i - \theta_y)|y_i - \theta_y|]$$

for $i = 1, 2, \ldots, n$. Thus, in considering the possible sample values produced by the 2^n sign combinations applied to these vector observations we note that it is only the 'sign' values that are affected. In other words, we may regard the conditioning as being on the magnitude $|x_i - \theta_x|, |y_i - \theta_y|, i = 1, 2, \ldots, n$, and also on the positions in the observed sequence where $\text{sgn}\,(x_i - \theta_x) \neq \text{sgn}\,(y_i - \theta_y)$.

Now we can see that various test statistics, other than those of Section 7.3.2, can be generated by transformations of the magnitudes $|x_i - \theta_x|, |y_i - \theta_y|$, in the style of similar methods discussed in Chapter 2. The vectors $(x_i - \theta_x, y_i - \theta_y)$ can be thought of as being replaced by

$$[\text{sgn}\,(x_i - \theta_x)\,T(|x_i - \theta_x|), \text{sgn}\,(y_i - \theta_y)\,T(|y_i - \theta_y|)]$$
$$= [T_{xi}(\theta_x, \theta_y), T_{yi}(\theta_x, \theta_y)]$$

We now consider the statistic

$$[T_{x.}(\theta_x, \theta_y), T_{y.}(\theta_x, \theta_y)] = [\sum T_{xi}(\theta_x, \theta_y), \sum T_{yi}(\theta_x, \theta_y)]$$

and its distribution under our permutation scheme. Clearly, the joint distribution of $T_{x.}$ and $T_{y.}$ can be tabulated quite simply in the same way as that of $X.$ and $Y.$. Further, the second-order moments of $T_{x.}$ and $T_{y.}$ are obtained like those of $X.$ and $Y.$ simply by replacing $|x_i - \theta_x|$ by $T(|x_i - \theta_x|)$, etc. Thus we find that

$$E\{T_{x.}(\theta_x, \theta_y)\} = E\{T_{y.}(\theta_x, \theta_y)\} = 0$$
$$\operatorname{var}\{T_{x.}(\theta_x, \theta_y)\} = \sum \{T(|x_i - \theta_x|)\}^2$$
$$\operatorname{var}\{T_{y.}(\theta_x, \theta_y)\} = \sum \{T(|y_i - \theta_y|)\}^2$$
$$\operatorname{cov}\{T_{y.}(\theta_x, \theta_y), T_{y.}(\theta_x, \theta_y)\} = \sum T_{xi}(\theta_x, \theta_y)T_{yi}(\theta_x, \theta_y)$$

this covariance can also be expressed as

$$\sum_P (T(|x_i - \theta_x|)T(|y_i - \theta_y|) - \sum_Q T(|x_i - \theta_x|)T(|y_i - \theta_y|)$$

where P denotes the subset of observations where $\operatorname{sgn}(x_i - \theta_x) = \operatorname{sgn}(y_i - \theta_y)$ and Q is its complement.

For testing a hypothesis specifying a value of (θ_x, θ_y) we shall use the counterpart of Q defined in (7.1), namely

$$Q_T(\theta_x, \theta_y) = [T_{x.}(\theta_x, \theta_y), T_{y.}(\theta_x \theta_y)]V_T^{-1}(\theta_x, \theta_y)$$
$$\times [T_{x.}(\theta_x \theta_y), T_{y.}(\theta_x, \theta_y)]^T \qquad (7.3)$$

where $V_T(\theta_x, \theta_y)$ is the 2×2 covariance matrix of $T_{x.}, T_{y.}$ with elements as given above. We have $E\{Q_T(\theta_x, \theta_y)\} = 2$, and under suitable conditions its conditional distribution is approximately χ_2^2.

Point estimates of θ_x and θ_y will be obtained by solving the equations

$$\sum \operatorname{sgn}(x_i - t_x)T(|x_i - t_x|) = 0$$
$$\sum \operatorname{sgn}(y_i - t_y)T(|y_i - t_y|) = 0$$

the estimates obtained in this way, $\hat{\theta}_x, \hat{\theta}_y$, are just those that would be obtained by treating the x and y values separately as univariate samples. Generally, of course, $\hat{\theta}_x$ and $\hat{\theta}_y$ are not independent.

7.3.4 Symmetric distributions: sign statistics

The simplest useful transformation is $T(u) = 1$. An effect of this transformation is that we essentially condition only on the total number of $(x_i - \theta_x, y_i - \theta_y)$ pairs in which $\text{sgn}(x_i - \theta_x) \neq \text{sgn}(y_i - \theta_y)$; this is essentially the same as the conditioning used in Section 7.3.1.

In the notation of Section 7.3.1, omitting the arguments (t_x, t_y) or (θ_x, θ_y), we see that

$$S_x = T_{x.} = N_{11} + N_{10} - N_{01} - N_{00} = n - 2N_{0.}$$
$$S_y = T_{y.} = N_{11} + N_{01} - N_{10} - N_{00} = n - 2N_{.0}$$

so that we are led back to the median tests of Section 7.3.1. In fact, following calculations according to Sections 7.2 and 7.1, we see that

$$\text{var}(S_x) = n = \text{var}(S_y)$$
$$\text{cov}(S_x, S_y) = n_{11} + n_{00} - n_{01} - n_{10}$$

and it is simple exercise to show that Q_S given by (7.3) becomes $(n_{11} - n_{00})^2/(n_{11} + n_{00}) + (n_{01} - n_{10})^2/(n_{01} + n_{10})$, which is the expression for Q given in Section 7.3.1.

An interesting difference between univariate and bivariate sign statistics should be noted. In the univariate case the conditional null distribution of the sign statistic is invariant with respect to the realized observations. Hence it is also the unconditional distribution of the statistic. However, in the bivariate case the conditional joint distribution of the two sign statistics depends on the sample configuration, specifically, on the value of $n_{00} + n_{11}$. The unconditional joint distribution is, therefore, a mixture of distributions which will depend on the true joint distribution of X and Y.

7.3.5 Symmetric distributions: rank statistics

Let $T(|x_i - \theta_x|) = \text{Rank}(|x_i - \theta_x|)$, that is, the rank of $|x_i - \theta_x|$ in the sequence $|x_j - \theta_x|, j = 1, 2, \ldots, n$, and similarly, let $T(|y_i - \theta_y|) = \text{Rank}(|y_i - \theta_y|)$. We shall use the notation W_x for $T_{x.}$, W_y for $T_{y.}$; W_x and W_y are sums of signed x and y ranks respectively and may be regarded as bivariate versions of the Wilcoxon signed rank statistic.

Here we have

$$\text{var}\{W_x(\theta_x, \theta_y)\} = \text{var}\{W_y(\theta_x, \theta_y)\} = n(n + 1)(2n + 1)/6$$

invariant with respect to the realized sample. However, $\text{cov}(R_x, R_y)$ does depend on the sample configuration. Transformation to ranks has, therefore, not produced statistics that are unconditionally distribution-free. A similar phenomenon was noted in connection with the sign statistics in Section 7.3.4.

Testing $(\theta_x, \theta_y) = (\theta_x^\circ, \theta_y^\circ)$

We use the statistic Q_R, being Q_T given by (7.3) with an obvious change in notation. The conditions for asymptotic joint normality of R_x and R_y, and for the asymptotic χ_2^2 distribution of Q_R hold here; see comments following Theorem 1.8.3. In principle it is relatively easy to evaluate the conditional distribution of Q_R exactly for small n. However, while it is possible it is not practical to prepare tables of this conditional distribution for all reasonably small values of n.

Example 7.6 Refer to the data of Example 7.3 and consider the hypothesis $H_0 : (\theta_x, \theta_y) = (-0.3, -0.7)$. The observed $(x_i - \theta_x, y_i - \theta_y)$ values are shown overleaf with their signed ranks.

From this table of ranks we obtain

$$\text{Observed} \quad W_x = -75$$
$$W_y = 121$$

$$\text{var}(W_x) = \text{var}(W_y) = 9455$$
$$\text{cov}(W_x, W_y) = 5824$$

$$V_T^{-1} = \begin{bmatrix} 9455 & -5824 \\ -5824 & 9455 \end{bmatrix}(55\,478\,049)^{-1}$$
$$Q_T = 5.36$$

At the 20% level, for which the results of Example 7.3 are relevant, we reject H_0. In other words, if we constructed an 80% confidence region for (θ_x, θ_y) using Q_T, the point $(-0.3, -0.7)$ would not be in it.

Confidence regions

A confidence region based on the rank statistics can be constructed by the usual inverse hypothesis-testing procedure. For a set of data such as that used in Example 7.6 the calculations are somewhat tedious. It is easy to write down the finite set of t_x and t_y values at which $W_x(t_x, t_y)$ and $W_y(t_x, t_y)$ change as t_x and t_y vary; they are the sets of pairwise averages of the x's and y's. However, it seems that actually performing the ranking and other calculations exhibited in Example 7.6 is unavoidable for at least some (t_x, t_y) points.

Point estimation

The point estimates of θ_x and θ_y are the respective Hodges–Lehmann estimates calculated from the x and y values, hence their large-sample

$x_i - \theta_x$	$y_i - \theta_y$	Ranks		$x_i - \theta_x$	$y_i - \theta_y$	Ranks		$x_i - \theta_x$	$y_i - \theta_y$	Ranks	
0.748	0.265	19	7	−0.388	0.032	−14	1	0.127	0.078	6	3
0.238	−0.131	11	−5	0.101	−0.329	5	−12	−0.344	−0.582	−13	−16
0.541	1.013	16	22	−0.128	0.793	−7	18	−0.287	0.327	−12	10
−1.881	−1.795	−28	−27	−2.018	−2.007	−29	−29	−1.241	0.499	−25	15
−0.156	0.200	−8	6	0.223	0.592	10	17	−0.956	0.869	−23	20
1.562	2.095	27	30	−0.687	−1.002	−18	−21	0.842	1.712	21	26
0.685	0.485	17	14	0.190	0.473	9	13	−0.488	−0.278	−15	−8
0.093	1.285	4	24	−2.223	−1.815	−30	−28	1.519	1.182	26	23
−0.885	−0.328	−22	−11	0.814	1.472	20	25	0.024	−0.119	1	−4
−1.020	−0.046	−24	−2	−0.063	−0.314	−2	−9	0.087	0.803	3	19

variances are as given in Section 2.2. To find the large-sample covariance of the estimates we need the expectation of the conditional $\text{cov}(W_x, W_y)$, which depends on the underlying joint distribution of X and Y.

A large n expression for $\text{cov}(W_x, W_y)$ can be obtained by using Lemma 2.1, Section 2.3.4, and making the substitutions

$$F_x(X_i) - \tfrac{1}{2} \quad \text{for} \quad \text{sgn}(X_i) \, \text{Rank}(|X_i|)/(n+1)$$
$$F_y(Y_i) - \tfrac{1}{2} \quad \text{for} \quad \text{sgn}(Y_i) \, \text{Rank}(|Y_i|)/(n+1)$$

$i = 1, 2, \ldots, n$; where F_x and F_y are the marginal distribution functions of X and Y, respectively. Then $\text{cov}(W_x, W_y) \simeq n^3 \omega_{xy}$, where

$$\omega_{xy} = \iint (F_x(u) - \tfrac{1}{2})(F_y(v) - \tfrac{1}{2}) f_{xy}(u, v) \, du \, dv$$

where f_{xy} is the joint density function of X and Y.

For large n we have from Section 2.3.3

$$[\partial E\{W_x(t_x, t_y)\}/\partial t_x]_{\mathbf{t}=\boldsymbol{\theta}} \simeq -2n^2 \int f_x^2(u) du = -2n^2 \bar{f}_u$$

$$[\partial E\{W_x(t_x, t_y)\}/\partial t_y]_{\mathbf{t}=\boldsymbol{\theta}} = 0$$

with similar expressions for W_y, where f_x and f_y denote the marginal densities of X and Y. Using the formula (1.11) the large sample covariance matrix of the Hodges–Lehmann estimates $\hat{\theta}_x, \hat{\theta}_y$ is approximately

$$\frac{1}{12n}\begin{bmatrix} 1/\bar{f}_x^2 & 3\omega_{xy}/\bar{f}_x\bar{f}_y \\ 3\omega_{xy}/\bar{f}_x\bar{f}_y & 1/\bar{f}_y^2 \end{bmatrix}$$

Efficacy

Using the results from above, the efficacy of the signed rank statistic (W_x, W_y) can be expressed as

$$\{\text{Efficacy}(\mathbf{W})\}^2 = 12\{\bar{f}_x^2 + \bar{f}_y^2 + 6\omega_{xy}\bar{f}_x\bar{f}_y\}/\{1 - 9\omega_{xy}^2\}$$

Example 7.7 Suppose that the joint distribution of X and Y is normal with $E(X) = E(Y) = 0$, $\text{var}(X) = \text{var}(Y) = 1$ and correlation coefficient ρ. Write

$$\rho_W = 3\omega_{xy}$$

for the correlation coefficient of the signed ranks. Then since $f_x =$

$$\bar{f}_y = 1/(2\sqrt{\pi})$$

$$\{\text{Efficacy}(\mathbf{W})\}^2 = (3/\pi)(2/(1 + \rho_W))$$

and using the result from Example 7.4

$$\{\text{Efficacy}(\mathbf{W})\}^2/\{\text{Efficacy (mean)}\}^2 = (3/\pi)(1 + \rho)/(1 + \rho_W).$$

The value of ρ_W is given by

$$\rho_W = (6/\pi)\sin^{-1}(\rho/2)$$

Puri and Sen (1971) p. 176.

In Example 7.7 a simple expression is given for ρ_W, but generally ρ_W, and π_{00} in Section 7.3.1 have to be obtained by numerical integration.

7.3.6 *Symmetric distributions: scores based on ranks*

We refer briefly to transformations of a type discussed in earlier chapters, namely $G\{R_i/(n + 1)\}$, where R_i is a rank and G is an inverse distribution function; the best known of these is perhaps $G = \Phi_*^{-1}$ where Φ_* is the standard half-normal distribution. The statistics (T_x, T_y) now become

$$G_x(\theta_1, \theta_2) = \sum \text{sgn}(x_i - \theta_x) G\{R_{xi}(\theta_x)/(n + 1)\} = \sum G_{xi}$$
$$G_y(\theta_1, \theta_2) = \sum \text{sgn}(y_i - \theta_y) G\{R_{yi}(\theta_y)/(n + 1) = \sum G_{yi}$$

where $R_{wi}(\theta_w) = \text{Rank}(|w_i - \theta_w|), w = x, y$

The conditional joint distribution of G_x and G_y, and of a quadratic form of the type Q_T, can be listed using the same principles as before, and the conditional second-order moments of G_x and G_y are as given for T_x and T_y with the appropriate substitutions. The following example illustrates the procedure for testing a hypothesis about (θ_x, θ_y) using G_x and G_y.

Example 7.8 We use the data of Example 7.6 but to save space show only the first few transformed ranks. The other results reported are for the full set of $n = 30$ observations. The transformation used is $G = \Phi_*^{-1}$ where $\Phi_*(u) = 2\{\Phi(u) - \frac{1}{2}\}$; Φ is the standard normal distribution function. Thus rank 19 becomes

$$\Phi_*^{-1}(19/31) = 0.865$$

Signed ranks		Signed transformed ranks	
19	7	0.865	0.287
11	− 5	0.460	− 0.203
16	22	0.700	1.057
− 28	− 27	− 1.661	− 1.518

etc.

$$G_x = 4.993 \qquad G_y = 5.754$$
$$\sum G_{xi}^2 = 27.083\,382 = \sum G_{yi}^2$$
$$\sum G_{xi} G_{yi} = 18.505\,088$$

$Q_G = Q_T = 6.80$

The results above give G_x and G_y for the hypothetical $(\theta_x, \theta_y) = (-0.3, -0.7)$; see also Example 7.6. Treating $Q_T = 6.80$ as an observation on a χ_2^2 random variable, the hypothesis $H_0 : (\theta_x, \theta_y) = (-0.3, -0.7)$ is rejected at the 20% level.

The point estimates of θ_1 and θ_2 yielded by solving the estimating equations

$$G_x(t_1, t_2) = 0, \quad G_y(t_1, t_2) = 0$$

are just the estimates discussed in Section 2.2. The efficacy of the test, based on Q_G, can be derived using results on $(\partial E\{G_x(t_1, t_2)\}/\partial t_1)_{t=\theta}$ from Chapter 2 and methods similar to those of Section 7.3.5 for the unconditional covariance of G_x and G_y. Details will not be given.

7.3.7 Symmetric distributions: robust transformations

Let $\psi(u)$ be one of the transformations associated with M-estimates; typically $\psi(u)$ is a monotone, continuous, and differentiable function of u. Now we put

$$M_{\psi x}(\theta_x, \theta_y) = \sum \text{sgn}\,(x_i - \theta_x)\psi(|x_i - \theta_x|)$$
$$M_{\psi y}(\theta_x, \theta_y) = \sum \text{sgn}\,(y_i - \theta_y)\psi(|y_i - \theta_y|)$$

and apply the permutational methods outlined above. We obtain an exact conditional test procedure based on the M-statistics and find an exact confidence region for (θ_x, θ_y).

Since $M_{\psi x}$ and $M_{\psi y}$ are sums of independent random variables, it is somewhat easier to write down expressions for the relevant efficacy and the large-sample covariance matrix of the point estimate. If $\psi(u)$ is 'odd', i.e., such that $\psi(-u) = -\psi(u)$, we can simply write

$$M_{\psi x}(\theta_x, \theta_y) = \sum \psi(x_i - \theta_x)$$

and $M_{\psi y}$ similarly. Then conditionally,

$$\text{var}\,(M_{\psi x}) = \sum \{\psi(x_i - \theta_x)\}^2$$
$$\text{var}\,(M_{\psi y}) = \sum \{\psi(y_i - \theta_y)\}^2$$
$$\text{cov}\,(M_{\psi x}, M_{\psi y}) = \sum \psi(x_i - \theta_x)\psi(y_i - \theta_y)$$

At $(\theta_x, \theta_y) = (0, 0)$ the unconditional versions of these moments are

$$n\int \psi^2(u)f_x(u)\mathrm{d}u, \quad n\int \psi_y^2(u)f_y(u)\mathrm{d}u, \quad n\int\int \psi(x)\psi(y)f(x, y)\mathrm{d}x\,\mathrm{d}y$$

With suitable choice of ψ, normal approximation of the joint distribution of the M-statistics will be possible. This makes it relatively easy to test a hypothesis about (θ_x, θ_y) through the appropriate version of Q_T. Finding a joint confidence region is a fairly straightforward computational job because we need find only the (t_x, t_y) values such that

$$\{M_{\psi x}(t_x, t_y), M_{\psi y}(t_x, t_y)\}\, V_\psi^{-1}(t_1, t_2)\, \{M_{\psi x}(t_x, t_y), M_{\psi y}(t_x, t_y)\}' \leqslant C$$

where the elements of $V_\psi(t_x, t_y)$ are given by the formulae for the moments of the M-statistics as given above with (t_x, t_y) replacing (θ_x, θ_y).

7.3.8 Functions of θ_x and θ_y

We need to investigate joint inference for θ_x and θ_y, in particular joint estimation of (θ_x, θ_y), because a parameter of interest may be a function $A(\theta_x, \theta_y)$ of θ_x and θ_y. The simplest non-trivial function is linear, say $a_x\theta_x + a_y\theta_y$, with a_x and $b_x \neq 0$. The case $a_x = 1, a_y = 0$ (or $a_x = 0, a_y = 1$) is of course also of interest, but in the location model that we are considering it reduces to inference about the location of the marginal X-distribution. Note that this situation differs somewhat from some univariate problems involving two parameters. Now, while exact joint inference about (θ_x, θ_y) is possible, the problem of exact inference about $A(\theta_x, \theta_y)$ generally appears insoluble. This is a problem of nuisance parameters which we have encountered before.

Statistically approximate inference about $A(\theta_x, \theta_y)$ is possible by using the large-sample covariance matrix of the estimates of θ_x and θ_y; alternatively, exact but conservative inference can be made, using in essence an exact joint confidence region for (θ_x, θ_y). Consider, for example, estimation of $\Delta = \theta_y - \theta_x$. Suppose that a joint 80% confidence region for (θ_x, θ_y) has been graphed with θ_y as ordinate and θ_x as abscissa. Then two straight lines of slope 1 drawn as tangents to the confidence region have θ_y intercepts Δ_L, Δ_u and the integral (Δ_L, Δ_u) is a confidence interval for Δ with confidence coefficient at least 80%.

Alternatively, if approximate normality of the estimates $\hat{\theta}_x$ and $\hat{\theta}_y$ is assumed, $\hat{\Delta} = \hat{\theta}_y - \hat{\theta}_x$ has approximate variance given by the appropriate large-sample covariance matrix of $(\hat{\theta}_x, \hat{\theta}_y)$. However, a problem remains in that the elements of this matrix have, themselves, to be estimated. Sample values of the derivatives that enter into their calculation can be used, but obtaining good estimates of these elements is, as yet, not a fully resolved matter.

7.4 Two-sample location problems

7.4.1 *Introduction: randomization*

Two bivariate random samples are considered, one of size m, the other of size n. In the 'm-sample' we have (x_{i1}, y_{i1}), $i = 1, 2, \ldots, m$, and in the 'n-sample', (x_{j2}, y_{j2}), $j = 1, 2, \ldots, n$. These samples are assumed to be drawn from populations with distribution functions $F_1(x, y)$ and $F_2(x, y)$. Tests of equality of these two distribution functions may be considered in a rather general context, but we shall, as in Chapter 4, make some simplifying assumptions about the alternative hypotheses, thus narrowing down the classes of test statistics to be considered.

Much the simplest type of alternative seems to be the location-shift alternative, $F_2(x, y) = F_1(x + \theta_x, y + \theta_y)$. Under such an alternative the sequence of $N = m + n$ values

$$(x_{i1}, y_{i1}), i = 1, 2, \ldots, m, (x_{j2} - \theta_x, y_{j2} - \theta_y), j = 1, 2, \ldots, n$$

can be considered as having been drawn at random from the same population. For convenience we shall refer to the above sequence as $(u_i(\boldsymbol{\theta}), v_i(\boldsymbol{\theta}))$, $i = 1, 2, \ldots, N$, so that for example $u_1(\boldsymbol{\theta}) = x_{11}, u_{m-1}(\boldsymbol{\theta}) = x_{12} - \theta_x$, etc. The basic randomization argument, as before, uses conditioning on the observed set of (u, v) values and we consider the conditional distributions of various statistics under a scheme of random partitioning the N values of (u, v) into subsets of sizes m and n.

7.4.2 *Medians and sign tests*

Let $\hat{u}$ be the median of the N values u_i, and $\hat{v}$ the median of the v_i values and let $N_{00} = \#(u_i \leqslant \hat{u}, v_i \leqslant \hat{v})$,

$$N_{01} = \#(u_i \leqslant \hat{u}, v_i > \hat{v}), N_{10} = \#(u_i > \hat{u}, v_i \leqslant \hat{v})$$
$$N_{11} = \#(u_i > \hat{u}, v_i > \hat{v}). \text{ Also put } N_0. = N_{00} + N_{01},$$
$$N_1. = N_{10} + N_{11}, \text{ etc.}$$

If N is even we note that

$$N_0. = N_1. = N_{.0} = N_{.1} = N/2 \text{ and } N_{11} = N_{00}, N_{01} = N_{10}.$$

According to the randomization scheme discussed in Section 7.4.1 we select m pairs at random without replacement from the total of N pairs (u_i, v_i). Let $M_{00}(\theta_x, \theta_y) = \#(\text{selected pairs such that } u_i \leqslant \hat{u}, v_i < \hat{v})$, with M_{01}, M_{10}, M_{11} defined similarly. Under this randomization scheme,

$$E(M_1.) = mN_1./N = m/2 \qquad (N \text{ even})$$
$$E(M_{.1}) = mN_{.1}/N = m/2 \qquad (N \text{ even})$$

Therefore we base a test of a specified (θ_x, θ_y) on the bivariate statistic $(M_1., M_{.1})$ which can be thought of as a median test.

In order to use the conditional randomization scheme outlined in Section 7.4.1 for exact median tests note that the null hypothesis need not specify that F_1 and F_2 are identical. In fact, we only need $F_1(\theta_x, \theta_y) = F_2(\theta_x, \theta_y)$; but we shall concentrate on the location-difference case.

The joint conditional distribution of M_{00}, M_{01}, M_{10} (and $M_{11} = m - M_{00} - M_{10} - M_{01}$) is a multivariate hypergeometric distribution with parameters $(N, m, N_{00}, N_{01}, N_{11})$, whose probability function is easily written down. The following formulae are readily obtainable:

$$\text{var}(M_1.) = m \frac{N_1.(N - N_1.)}{N} \left(\frac{N - m}{N - 1} \right)$$

$$\text{var}(M_{.1}) = m \frac{N_{.1}(N - N_{.1})}{N^2} \left(\frac{N - m}{N - 1} \right)$$

$$\text{cov}(M_1., M_{.1}) = m \left(\frac{N_{11}}{N} - \frac{N_1. N_{.1}}{N^2} \right) \left(\frac{N - m}{N - 1} \right)$$

Note that here again, despite the transformation to signs, the conditional joint distribution of $M_1.$ and $M_{.1}$ is not invariant with respect to the sample configuration.

Testing a specified (θ_x, θ_y)

Following precedents we shall base our hypothesis test on

$$Q_M = (M_1. - mN_1./N, M_{.1} - mN_{.1}/N) V_M^{-1}$$
$$\times (M_1. - mN_1./N, M_{.1} - mN_{.1}/N)'$$

where V_M is the covariance matrix of $M_{1.}$ and $M_{.1}$ with elements as given above; if N is even the formulae are simpler, with

$$\operatorname{var}(M_{1.}) - \operatorname{var}(M_{.1}) = (m/4)(N-m)/(N-1)$$
$$\operatorname{cov}(M_{1.}, M_{.1}) = m(N_{11}/N - 1/4)(N-m)/(N-1).$$

The exact distribution of Q_M can be evaluated; $E(Q_M) = 2$ and we shall take the distribution of Q_M to be approximately χ^2 for m and n moderately large.

Example 7.9

$$m = 3 : (-0.8, -0.7), (-0.6, +0.6), (+0.8, +0.9)$$
$$n = 3 : (-0.9, -0.3), (+0.5, +1.2), (+0.8, +0.7)$$
$$H_0 : (\theta_x, \theta_y) = (0.3, 1.0)$$

modified 'n-sample': $(-1.2, -1.3), (0.2, +0.2), (0.5, -0.3)$

In the pooled sample, $(\hat{u}, \hat{v}) = (-0.2, -0.05)$, $N_{00} = 2$, $N_{01} = 1$, $N_{10} = 1$, $N_{11} = 2$, and the joint distribution of $M_{1.}$, $M_{.1}$ is as follows:

		$M_{.1}$				
		0	1	2	3	
	0		1/20			1/20
$M_{1.}$	1	1/20	4/20	4/20		9/20
	2		4/20	4/20	1/20	9/20
	3		1/20			1/20
		1/20	9/20	9/20	1/20	20/20

From this table, and the formulae given above,

$$\operatorname{var}(M_{1.}) = \operatorname{var}(M_{.1}) = 9/20,$$
$$\operatorname{cov}(M_{1.}, M_{.1}) = 3/20,$$

and the distribution of Q_M is as follows:

q :	5/6	5/3	5
$20\Pr(Q_M = q)$:	8	8	4

This example only illustrates computational methods; the sample is too small for useful inference based on Q_M.

Confidence regions, point estimation
Finding a joint confidence region can be expedited somewhat by

noting, as we have described in Section 4.3.3, that $M_{1.}(t_x, t_y)$ is a step function of t_x with jumps occurring only at certain $x_{i1} - x_{i2}$ differences; $M_{.1}(t_x, t_y)$ behaves similarly. However, complications arise because N_{11} might also change as t_x or t_y is varied.

The point estimates of θ_x and θ_y are obtained by solving the estimating equations

$$M_{1.}(t_x, t_y) = 1/2; \quad M_{.1}(t_x, t_y) = 1/2 \tag{7.4}$$

and are readily seen to be just the respective univariate estimates obtained in Chapter 4. However, the two estimates are not independently distributed. To find their large-sample covariance matrix we need

$$E[\text{cov}\{M_{1.}(\boldsymbol{\theta}), M_{.1}(\boldsymbol{\theta})\}] = m\left(\frac{E(N_{11})}{N} - \frac{N_{1.}N_{.1}}{N^2}\right)\left(\frac{N-m}{N-1}\right)$$

For the present purpose we may take the medians of both F_1 and F_2 to be $(0,0)$, so that we can write

$$E(N_{11}) = E(N_{00}) = NF_1(0,0) = N\pi_{00}$$

If N is even, we then have

$$E[\text{cov}(M_{1.}(\boldsymbol{\theta}), M_{.1}(\boldsymbol{\theta})] = m(\pi_{00} - 1/4)(N-m)/(N-1)$$

Approximating terms $N - 1$ by N for large N, and using results from Chapter 4 for quantities like $(\partial E\{M_{1.}(t_x, t_y)\}/\partial t_x)_{t=\theta}$ the large-sample covariance matrix of the point estimates of θ_x and θ_y derived as solutions of (7.4) is

$$\left(\frac{N}{4mn}\right)\begin{bmatrix} 1/f_x^2 & (4\pi_{00} - 1)/f_x f_y \\ 4(\pi_{00} - 1)f_x f_y & 1/f_y^2 \end{bmatrix},$$

where f_x, f_y are the marginal common densities at the median values.

Efficacy
The efficacy of the test of (θ_x, θ_y) based on $(M_{1.}, M_{.1})$ can be calculated from the results given above; it is

$$4\{f_x^2 + f_y^2 + 2f_x f_y(1 - 4\pi_{00})\}/\{1 - (4\pi_{00} - 1)^2\}$$

7.4.3 *Mean statistics*

Let $\bar{u}(t), \bar{v}(t)$ be the means of the $u_i(t), v_i(t)$ values respectively. As an obvious generalization of the mean statistic of Chapter 4 we consider

$$A_x(\mathbf{t}) = \sum_{i=1}^{m} x_{1i} - m\bar{u}(\mathbf{t}), \quad A_y(\mathbf{t}) = \sum_{i=1}^{m} y_{1i} - m\bar{v}(\mathbf{t})$$

As we have indicated in Section 7.4.1, if $\mathbf{t} = \boldsymbol{\theta}$, all u_i, v_i values may be regarded as having been drawn from the same population and we can use the conditional permutation argument.

Let

$$\sigma_u^2 = (1/N)\sum(u_i - \bar{u})^2, \quad \sigma_v^2 = (1/N)\sum(v_i - \bar{v})^2 \qquad (7.5)$$

$$\sigma_{uv} = (1/N)\sum(u_i - \bar{u})(v_i - \bar{v}).$$

Then the first- and second-order moments of the joint permutation distribution of A_x and B_x are

$$E(A_x) = E(A_y) = 0$$

and

$$\text{var}(A_x) = m\sigma_u^2(N - m)/(N - 1)$$

$$\text{var}(A_y) = m\sigma_v^2(N - m)/(N - 1)$$

$$\text{cov}(A_x, A_y) = m\sigma_{uv}(N - m)/(N - 1)$$

If V_A is the covariance matrix of A_x and B_x, then the exact conditional permutation distribution of

$$Q_A = (A_x, A_y)V_A^{-1}(A_x, A_y)'$$

can be found; $E(Q_A) = 2$, and under suitable conditions the permutation distribution of Q_A will be approximately χ_2^2.

Hypothesis testing and confidence limits:
To test $H_0 : \boldsymbol{\theta} = \boldsymbol{\theta}°$ the calculations indicated by the formulae above are performed with $\boldsymbol{\theta}°$ replacing $\mathbf{t}$, and the resulting value of Q_A is compared with the appropriate χ_2^2 quantile. Confidence regions are found in the usual inverse hypothesis-testing manner. When a χ^2 approximation of the permutation distribution of Q_A is applicable, a $100\beta\%$ confidence region is given by the (t_x, t_y) set satisfying

$$Q_A(t_x, t_y) \leqslant \chi_2^2(\beta)$$

Example 7.10 We use the data of Example 7.9 and test $H_0 : (\theta_x, \theta_y) = (0.3, 1.0)$, as before. The collection of $N = 6$ (u_i, v_i) values are

$$(-0.8, -0.7), (-0.6, +0.6), (+0.8, +0.9)$$

$$(-1.2, -1.3), (+0.2, +0.2), (+0.5, -0.3)$$

giving

$$\bar{u} = 0.183\,333 \qquad\qquad \bar{v} = -0.100\,000$$

$$\sigma_u^2 = 0.528\,055 \qquad\qquad \sigma_v^2 = 0.570\,000$$

$$\sigma_{uv} = 0.376\,667$$

$$A_x = -0.6 - 3\bar{u} = -0.050\,000$$

$$A_y = +0.8 - 3\bar{u} = +1.100\,000$$

$$V_A^{-1} = \frac{5}{9}\begin{bmatrix} 3.5823 & -2.3673 \\ -2.3673 & 3.3187 \end{bmatrix}$$

$$Q_A = 2.3805$$

The exact null distribution of Q_A for this case has been enumerated. There are 10 distinct Q_A values occurring with equal probability; they are:

$$0.1702, \quad 0.6420, \quad 0.6530, \quad 1.4558, \quad 1.7155,$$

$$1.8197, \quad 2.3261, \quad 2.3805, \quad 4.1734, \quad 4.6632$$

Thus $\Pr(Q_A \geqslant 2.3805) = 3/10$.

An exact confidence region, based on the mean statistics, can be found but the calculations are lengthy because the exact distribution of Q_A has to be listed for every trial (t_x, t_y). Using the χ_2^2 approximation indicated above leads to relatively simple calculations.

Point estimation

The point estimates obtained by solving

$$A_x(\mathbf{t}) = 0, \quad A_y(\mathbf{t}) = 0$$

are simply the usual differences of means; thus

$$(\theta_x, \theta_y) \text{ is estimated by } (x_{2.} - x_{1.}, y_{2.} - y_{1.})$$

where $x_{1.} = (1/M)\sum_{i=1}^{m} x_{1i}$, etc.

7.4.4 *Rank statistics*

The rank statistics which we shall consider are obtained by ranking the N values of u_i and also the N values of v_i and then using the sums of the ranks in the M-sample to define two test statistics:

$$W_x(\mathbf{t}) = \sum_{i=1}^{m} \text{Rank}\,\{u_i(\mathbf{t})\} - m(N+1)/2$$

$$W_y(\mathbf{t}) = \sum_{i=1}^{m} \text{Rank}\,\{v_i(\mathbf{t})\} - m(N+1)/2$$

The two statistics are the Wilcoxon two-sample rank sum statistics defined for the x and y values.

Under our permutation scheme we have

$$E(W_x) = E(W_y) = 0$$
$$\text{var}(W_x) = \text{var}(W_y) = mn(N+1)/12$$
$$\text{cov}(W_x, W_y) = m\sigma_{uv}(R)(N-m)/(N-1)$$

where $\sigma_{uv}(R)$ is calculated according to the formula for σ_{uv} in (7.5) but with u_i and v_i replaced by their ranks. The value of $\sigma_{uv}(R)$ will depend on the sample configuration and on $\mathbf{t}$; this is a symptom of the joint permutation distribution of W_x and W_y not being invariant with respect to the realized samples. Here, as in Section 7.4.3, a conditionally exact distribution-free procedure is not unconditionally distribution-free when the observations are transformed.

The quadratic form used for testing a hypothesis about (θ_x, θ_y) is

$$Q_W = (W_x, W_y) V_W^{-1}(W_x, W_y)'$$

where V_W is the conditional covariance matrix of (W_x, W_y) with elements as defined above. For large n its distribution is approximately χ_2^2

Example 7.11 With the data of Example 7.10 and the trial $(\theta_x, \theta_y) = (0.3, 1.0)$ as before, the set of pairs of ranks is

$$(2,2), \ (3,5), \ (6,6), \ (1,1), \ (4,4), \ (5,3)$$

giving $\sigma_{uv}(R) = 27/12$,

$$V_W^{-1} = \left(\frac{60}{4464}\right)\begin{bmatrix} 35 & -27 \\ -27 & 35 \end{bmatrix}$$

and observed $Q_W = 2.1505$.

In the exact null distribution of Q_W there are the following six distinct values of Q_W with probabilities shown in parentheses.

0.0538 (1/10), 0.4839 (2/10), 1.7204 (2/10), 2.1505 (2/10),

2.6344 (1/10), 4.3011 (2/10)

Changing the trial (θ_x, θ_y) to $(0.3, 0.0)$ changes the set of ranks to

$$(2,1), \ (3,3), \ (6,5), \ (1,2), \ (4,6), \ (5,4)$$

giving $\sigma_{uv}(R) = 29/12$, a new observed Q_W, and a new conditional distribution of Q_W.

Confidence region, point estimation
A joint confidence region can be obtained in the usual way, but may

be somewhat tedious to compute. Note that the only (t_x, t_y) points for which computations may have to be performed are those with at least one of W_x or W_y 'jumps'; these are discussed in Section 4.3.2.

The point estimates, also, are those discussed in Section 4.3.4, and the only new element here is that the two point estimates are correlated. To find their large-sample covariance we require $E(\sigma_{uv}(R))$. Without loss of generality, for the present purpose, we take $F_1(x, y) = F_2(x, y) = F(x_1, y)$ with corresponding densities $f(u, v)$. Now

$$\sigma_{uv} = (1/N)\sum R_{xi}R_{yi} - (N + 1)^2/4$$

where (R_{xi}, R_{yi}) are the ranks of the X and Y values in the N independent pairs $(X_i, Y_i, i = 1, 2, \ldots, N$. Using methods of earlier chapters, for example Lemma 2.1, Section 2.3.4., we have as a large-N approximation

$$E(R_{xi}R_{yi}) \simeq N^2 \int\int F(u, \infty)F(\infty, v)f(u, v)\,du\,dv$$

The values of $[\partial\{W_z(t)\}/\partial t_z]_{t=\theta}$, $z = x, y$, that are also required for the large-sample covariance of the point estimates are given in Section 4.3.4; they are $mn\bar{f}_x, mn\bar{f}_y$, where f_x, f_y are the marginal X, Y densities and $\bar{f}_x = \int f_x^2(u)\,du, f_y = \int \bar{f}_y^2(v)\,dv$.

7.4.5 Rank-based scores and other transformations

Tests like those described in Sections 7.4.3 and 7.4.4 can be performed after transforming the (u_i, v_i) values to rank-based scores of the type $G\{\text{Rank}\,(u_i)/(N + 1)\}$, etc., where typically G is an inverse distribution function. Similarly, robust transformations of the kind associated with M-estimates can be used. By either method the basic randomization argument gives an exact distribution-free test and will produce an exact joint confidence region for (θ_x, θ_y).

Point estimates are obtained by solving the estimating equations obtained by the simple application of the method of moments that we have been using throughout. Details of large-sample variances and covariances are very similar to others that have already been presented, and will be left as exercises.

Example 7.12 This example illustrates the calculations based on scores of the type $\Phi^{-1}\{\text{Rank}/(N + 1)\}$ for a moderately large set of data. They are taken from an example in Morrison (1967, p. 154), and are the results of a drug trial on mice, which were divided at random into 'Control' and 'Drug' groups. The levels of three biochemical compounds found in the brain were the response variables of interest. In the following list we report only the results of assays for two of these compounds.

	Control				Drug		
$A(x)$	Score	$B(y)$	Score	$A(x)$	Score	$B(y)$	Score
1.21	0.512	0.61	1.711	1.40	1.711	0.50	0.857
0.99	− 0.939	0.43	0.055	1.17	0.109	0.39	− 0.512
0.80	− 1.711	0.35	− 0.857	1.23	0.781	0.44	0.164
0.85	− 1.359	0.48	0.641	1.19	0.391	0.37	− 0.641
0.98	− 1.125	0.42	− 0.109	1.38	1.359	0.42	− 0.109
1.15	− 0.055	0.52	1.125	1.17	0.109	0.45	0.333
1.10	− 0.276	0.50	0.857	1.31	1.125	0.41	− 0.276
1.02	− 0.641	0.53	1.359	1.30	0.939	0.47	0.512
1.18	0.276	0.45	0.333	1.22	0.641	0.29	− 1.359
1.09	0.451	0.40	− 0.391	1.00	− 0.781	0.30	− 1.125
				1.12	− 0.164	0.27	1.711
				1.09	− 0.451	0.35	− 0.857

The null hypothesis under test is that the drug had no effect on the levels of A or B; $(\theta_x, \theta_y) = (0, 0)$.

The rank of the A observation 1.21 in the pooled sample is 16 and its transformed value is $\Phi^{-1}(16/23) = 0.512$ from a table of the standard normal distribution. We also illustrate a standard method of dealing with ties: the two tied A values at 1.09 are given the mean, $7\frac{1}{2}$, of their ranks 7 and 8. This method of dealing with tied ranks introduces little extra computational work because the covariance always has to be calculated for every sample.

Denoting the A and B scores by $S_i(A)$, $S_i(B)$, $S_.(A) = (1/N)\sum S_i(A)$; $\sigma^2(A) = \sum S_i^2(A)/N - S_.^2(A)$; etc., we have from the table of scores

$$\sum_{i=1}^{N} S_i(A) = 0.000 \qquad\qquad \sum_{i=1}^{N} S_i(B) = 0.000$$

$$\sum_{i=1}^{N} S_i^2(A) = 16.913\,042 \qquad\qquad \sum S_i^2(B) = 16.868\,418$$

$$\sum_{i=1}^{N} S_i(A)S_i(B) = 2.661\,966$$

$$\sigma^2(A) = 0.768\,775 \qquad\qquad \sigma^2(B) = 0.766\,746$$

$$\sigma(A, B) = 0.120\,998$$

The observed values of the test statistics are the sums of scores of the control sample

$$A_c = \sum_{\text{control}} S_i(A) = -5.769, \quad B_c = \sum_{\text{control}} S_i(B) = 4.724$$

and using these to compute the usual quadratic form test statistic we obtain the observed value

$$Q = \frac{(N-1)}{mn}[A_c, B_c]\begin{bmatrix} \sigma^2(A) & \sigma(A, B) \\ \sigma(A, B) & \sigma^2(B) \end{bmatrix}^{-1}\begin{bmatrix} A_c \\ B_c \end{bmatrix} = 15.00$$

Treating Q as χ_2^2 distribution under the null hypothesis, the null hypothesis is clearly rejected.

7.5 Three-sample location problems

7.5.1 Randomization

By straightforward extension of the arguments and notation of Section 7.4.1, we now consider samples of sizes n_1, n_2, n_3 from three populations with distribution functions $F_1(x, y), F_2(x, y) = F_1(x + \theta_{x2}, y + \theta_{y2}), F_3(x, y) = F(x + \theta_{x3}, y + \theta_{y3})$. The samples from the three populations are $(x_{ri}, y_{ri}), r = 1, 2, 3, i = 1, 2, \ldots n_r$. We also define a sequence of $u_i(\theta)$ values as the sequence of $x_{11}, x_{12}, \ldots, x_{1n}, x_{21} - \theta_{x2}, \ldots, x_{2n_2} - \theta_{x2}, x_{31} - \theta_{x3}, \ldots x_{3n_3} - \theta_{x3}$; the sequence of $v_i(\theta)$ values is defined similarly using the y_{ri} values.

If the u_i and v_i values are obtained using the population θ values, the sequence $(u_i, v_i), i = 1, 2, \ldots, N$ can be regarded as a sample from the same bivariate population and the randomization scheme to be used is random partitioning of the N pairs into groups of sizes n_1, n_2, n_3. The exact conditional distribution of any test statistic or statistics can be obtained by listing all of the possible partitions.

7.5.2 The choice of test statistic

The question remaining is the choice of a suitable statistic for testing a hypothesis that specifies a set of θ values; note in passing that in the most common 'one-way analysis of variance' situation all of the θ values are specified as being zero. It is instructive to re-examine the test procedure suggested for the two-sample case, One may argue that the null hypothesis under test is acceptable if the observed mean point (x_1, y_1) of the m-sample is not 'too far' from the 'expected point' $(\bar{u}, \bar{v})$. If $\sigma_{uv}^2 = 0$ a natural measure of distance of (x_1, y_1) from

$(\bar{u}, \bar{v})$ would be $D = (x_{1.} - u)^2/\mathrm{var}(x_{1.}) + (y_{1.} - v)^2/\mathrm{var}(y_{1.})$. Since σ_{uv} is, in general, not zero, the distance is defined by the quadratic form Q_A given in Section 7.4.3; in the case $\sigma_{uv} = 0$, $Q_A = D$ as given above.

It is easy to transform the (u_i, v_i) values to (u_i^*, v_i^*) values such that, for this new sequence $\sigma_{uv}^* = 0$; the transformation is

$$\begin{bmatrix} u_i^* \\ v_i^* \end{bmatrix} = \begin{bmatrix} \sin A & \cos A \\ -\cos A & \sin A \end{bmatrix} \begin{bmatrix} u_i \\ v_i \end{bmatrix}$$

with A given by

$$-2\cos 2A\,\sigma_{uv} + \sin 2A(\sigma_v^2 - \sigma_u^2) = 0$$

Now we define $x_{1.}^*$ as the mean of u_i^* values belonging to the m-sample, and $y_{1.}^*$ similarly. Then

$$\begin{pmatrix} x_{r.}^* - \bar{u}^* \\ y_{r.}^* - \bar{v}^* \end{pmatrix} = \begin{bmatrix} \sin A & \cos A \\ -\cos A & \sin A \end{bmatrix} \begin{pmatrix} x_{r.} - \bar{u} \\ y_{r.} - \bar{v} \end{pmatrix}, \qquad r = 1, 2 \quad (7.6)$$

and

$$\begin{aligned} \sigma_u^{*2} &= \sin^2 A\,\sigma_u^2 + 2\sin A \cos A\,\sigma_{uv} + \cos^2 A\,\sigma_v^2 \\ \sigma_v^{*2} &= \cos^2 A\,\sigma_u^2 - 2\sin A \cos A\,\sigma_{uv} + \sin^2 A\,\sigma_v^2 \end{aligned} \qquad (7.7)$$

The distance of $(x_{1.}^*, y_{1.}^*)$ from $(\bar{u}^*, \bar{v}^*)$ is

$$\begin{aligned} D_1^* &= (x_{1.}^* - u^*)^2/\mathrm{var}(x_{1.}^*) + (y_{1.}^* - v^*)/\mathrm{var}(y_{1.}^*) \\ &= \frac{m(N-1)}{n}\left\{\left(\frac{x_{1.}^* - \bar{u}^*}{\sigma_u^*}\right)^2 + \left(\frac{y_{1.}^* - \bar{v}^*}{\sigma_v^*}\right)^2\right\} \end{aligned}$$

and in fact, $D_1^* = Q_A$.

In the three-sample case there are two mean points $(x_{1.}, y_{1.})$ and $(x_{2.}, y_{2.})$ whose distances from $(\bar{u}, \bar{v})$ have to be considered. If we first transform to (u_i^*, v_i^*) values as explained above, it is natural to consider using $D_1^* + D_2^*$ as a test statistic. However, in such an expression, although $x_{1.}^*$ is uncorrelated with $y_{1.}^*$ and $y_{2.}^*$, it is correlated with $x_{2.}^*$. Thus, while we can treat the $x_{r.}^*$ and $y_{r.}^*$ values 'independently', we should represent the sum of distances of $x_{1.}^*$ and $x_{2.}^*$ from $\bar{u}$ by a quadratic form, as in the one-way analysis of variance taking into account $\mathrm{cov}(x_{1.}^*, x_{2.}^*)$. The proposed test statistic then becomes

$$\begin{aligned} Q^* &= \left(\frac{N-1}{N}\right)\left(\frac{n_1 n_2}{n_3 \sigma_u^*}\right)[x_{1.}^* - \bar{u}^*, x_{2.}^* - \bar{u}^*]\begin{bmatrix} N/n_2 - 1 & 1 \\ 1 & N/n_1 - 1 \end{bmatrix}\begin{bmatrix} x_{1.}^* - \bar{u}^* \\ x_{2.}^* - \bar{u}^* \end{bmatrix} \\ &+ \left(\frac{N-1}{N}\right)\left(\frac{n_1 n_2}{n_3 \sigma_v^*}\right)[y_{1.}^* - \bar{v}^*, y_{2.}^* - \bar{v}^*]\begin{bmatrix} N/n_2 - 1 & 1 \\ 1 & N/n_1 - 1 \end{bmatrix}\begin{bmatrix} y_{1.}^* - \bar{v}^* \\ y_{2.}^* - \bar{v}^* \end{bmatrix} \end{aligned}$$

$$(7.8)$$

A measure identical to Q_A^* is obtained if a single vector of 4 elements,

$$\Delta' = (x_{1.} - \bar{u}, x_{2.} - \bar{u}, y_{1.} - \bar{v}, y_{2.} - \bar{v})$$

is formed whose elements have the 4×4 covariance matrix V_A, and then computing

$$Q_A = \Delta' V_A^{-1} \Delta = Q^*$$

The matrix V_A is

$$\left(\frac{1}{N-1}\right) \begin{bmatrix} \sigma_u^2 C_1 & -\sigma_u^2 & \sigma_{uv} C_1 & -\sigma_{uv} \\ -\sigma_u^2 & \sigma_u^2 C_2 & -\sigma_{uv} & \sigma_{uv} C_2 \\ \sigma_{uv} C_1 & -\sigma_{uv} & \sigma_v^2 C_1 & -\sigma_v^2 \\ -\sigma_{uv} & \sigma_{uv} C_2 & -\sigma_v^2 & \sigma_v^2 C_2 \end{bmatrix}$$

where $C_1 = N/n_1 - 1$, $C_2 = N/n_2 - 1$.

Finally it is instructive, and may be more efficient for computation, to note that the components due to x^* and y^* in Q^* are simply the values of the statistic B^* proposed for the one-way analysis of variance calculated from the $x^*(u^*)$ and $y^*(v^*)$ values respectively. Thus, if $B^*(u^*) = (N-1)B(u^*)/T(u^*)$, where $B(u^*)$, $T(u^*)$, are the between-group and within-group sums of squared differences for u^* values, with $B^*(v^*)$ defined similarly.

$$Q^* = B^*(u^*) + B^*(v^*) \tag{7.9}$$

See also Section 6.4.1.

7.5.3 Inference using Q_Δ (equivalently Q^*)

Under randomization the exact null distribution of Q^* can be tabulated. Its expectation is exactly 4, and under the conditions where the analogous quadratic forms that occur in univariate one-way analysis of variance have approximate χ^2 distributions, the randomization distribution of Q^* will be approximately χ_4^2.

Example 7.13 The following total of $N = 6$ observations is divided into three groups as shown

Sample 1 : $n_1 = 1$: $(-0.8, -0.7)$

Sample 2 : $n_2 = 2$: $(-0.6, +0.6), (+0.8, +0.9)$

Sample 3 : $n_3 = 3$: $(-1.2, -1.3), (+0.2, +0.2), (+0.5, -0.3)$

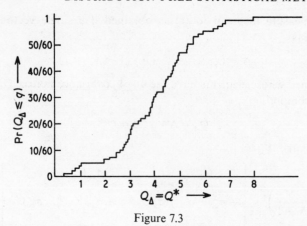

Figure 7.3

Suppose that we test the hypothesis H_0 that all θ values are zero, that is, the three samples come from the same population. Then the (u_i, v_i) values are simply the $N = 6$ vectors given above, giving

$$\bar{u} = -0.183\,33\dot{3}, \quad \bar{v} = -0.100\,000$$
$$\sigma_u^2 = 0.528\,05\dot{5}, \quad \sigma_v^2 = 0.570\,000, \quad \sigma_{uv} = 0.376\,66\dot{6}$$
$$\sin A = 0.681\,71, \quad \cos A = 0.726\,496$$

$$(x_{1.}\,; y_{1.}) = (-0.8, -0.7)$$
$$(x_{2.}, y_{2.}) = (+0.10, +0.75)$$

$$\bar{u}^* = -0.198\,733 \qquad \bar{v}^* = +0.644\,83$$
$$\sigma_u^* = 0.926\,256 \qquad \sigma_v^* = 0.171\,778$$

$$(x_{1.}^* - \bar{u}^*, y_{1.}^* - \bar{v}^*) = (-0.859\,653, \ 0.035\,753) \qquad \text{from (7.9)}$$
$$(x_{2.}^* - \bar{u}^*, y_{2.}^* - \bar{v}^*) = (0.812\,220, \ 0.372\,255)$$

$$Q^* = 4.4365$$

The value of Q^* has been evaluated for every one of the 60 possible partitions in this example, giving the conditional null distribution whose distribution function is depicted in Figure 7.3. From the tabulation of Q^* values,

$$\Pr\{Q^* \geqslant 4.4365\} = 25/60$$

7.5.4 Rank methods

By obvious extension of the rank methods for the two-sample case we now simply transform all the u_i values to their ranks, and do likewise for the v_i values. All calculations are now exactly like those outlined in Section 7.5.3 except that the original (u_i, v_i) are replaced by their ranks. Some simplifications occur because the values of $\bar{u}, \bar{v}, \sigma_u^2, \sigma_v^2$ become

$$\bar{u}(\text{Rank}) = \bar{v}(\text{Rank}) = (N + 1)/2$$
$$\sigma_u^2(\text{Rank}) = \sigma_v^2(\text{Rank}) = (N^2 - 1)/12$$

The value of $\sigma_{uv}(\text{Rank})$ will depend on the sample configuration, thus reminding us that, despite the transformation to ranks the exact test procedures for the θ's jointly are only conditionally distribution-free. However, if we use the method of calculation of distance as in equation (7.8) we find that $A = 45°$, giving

$$\bar{u}^*(\text{Rank}) = (N + 1)/\sqrt{2}$$
$$\bar{v}^*(\text{Rank}) = 0$$
$$\sigma_u^{*2}(\text{Rank}) = (N^2 - 1)/12 + \sigma_{uv}(\text{Rank})$$
$$\sigma_v^{*2}(\text{Rank}) = (N^2 - 1)/12 - \sigma_{uv}(\text{Rank})$$

We shall not rewrite the formulae of Section 7.5.3 in terms of ranks for our present purpose. Note that we shall use $x_{1.}$ to denote the mean of the ranks of u_i values of Sample 1, etc. The following example illustrates the method.

Example 7.14 We consider the problem of Example 7.13 using the rank transformation. The transformed values are

Sample 1 : (2, 2)
Sample 2 : (3, 5)   (6, 6)
Sample 3 : (1, 1)   (4, 4)   (5, 3)

$$(x_{1.} - \bar{u}, y_{1.} - \bar{v}) = (-1.5, -1.5)$$
$$(x_{2.} - \bar{u}, y_{2.} - \bar{v}) = (1.0, 2.0)$$

$$(x_{1.}^* - \bar{u}^*, y_{1.}^* - \bar{v}^*) = (-3.0/\sqrt{2}, 0)$$
$$(x_{2.}^* - \bar{u}^*, y_{2.}^* - \bar{v}^*) = (3.0/\sqrt{2}, 1.0/\sqrt{2})$$

$$\sigma_{uv} = 17/12$$
$$\sigma_u^* = 35/12 + 17/12 = 52/12$$
$$\sigma_v^* = 35/12 - 17/12 = 18/12$$

From (7.8),

$$Q^*(\text{Rank}) = \left(\frac{10}{18}\right)\left(\frac{12}{52}\right)\left(\frac{1}{2}\right)(-3.0, 3.0)\begin{bmatrix} 2 & 1 \\ 1 & 5 \end{bmatrix}\begin{bmatrix} -3.0 \\ 3.0 \end{bmatrix}$$

$$+ \left(\frac{10}{18}\right)\left(\frac{12}{18}\right)\left(\frac{1}{2}\right)(0, 1.0)\begin{bmatrix} 2 & 1 \\ 1 & 5 \end{bmatrix}\begin{bmatrix} 0 \\ 1.0 \end{bmatrix}$$

$$= 3.8105$$

This observed Q^* value is slightly smaller than expected value of 4, hence without further calculations we can accept H_0.

If an exact test at, say, the 5% level is required we need only find the 3 largest Q^* values, which are fairly easy to identify. They are given by the following 3 partitions of the total sample, giving the Q^* values shown:

$$(3, 5) \quad (2, 2), \ (1, 1) \quad (4, 4) \ (5, 3) \ (6, 6) \quad Q^* = 6.2251$$
$$(5, 3) \quad (2, 2), \ (1, 1) \quad (4, 4) \ (3, 5) \ (6, 6) \quad Q^* = 6.2251$$
$$(6, 6) \quad (1, 1), \ (2, 2) \quad (4, 4) \ (3, 5) \ (5, 3) \quad Q^* = 5.7692$$

7.5.5 Sign statistics and other transformations

Sign statistics are obtained on replacing u_i by $\text{sgn}(u_i - \hat{u})$, where $\hat{u}$ is the median of the Nu_i values, and the v_i values are transformed similarly. Again, the ensuing calculations are just those outlined in Section 7.5.3 with the original (u_i, v_i) values replaced by their transforms. We have

$$\bar{u}(\text{sgn}) = \bar{v}(\text{sgn}) = 0$$

$$\sigma_u^2(\text{sgn}) = \sigma_v^2(\text{sgn}) = \begin{cases} N & N \quad \text{even} \\ N-1, & N \quad \text{odd} \end{cases}$$

but $\sigma_{uv}(\text{sgn})$ will depend on the sample configuration.

Further details will not be given except to note that the method, if applied to the two-sample case, leads to the same analysis as that described in Section 7.4.2 when n is even. Without change, the approach of Section 7.5.3 also copes with n odd.

Using other transformation based on ranks or robust transformations is straightforward, as far as testing of a joint null hypothesis is concerned. The steps of Section 7.5.3 are followed after applying the relevant transformation to (u_i, v_i) values.

7.6 Multiple-sample location problems

The randomization argument here is a straightforward extension of that used in the two- and three-sample cases; if there are k samples of sizes $n_1, n_2, \ldots, n_k$, summing to N, we consider partitions of the N vectors (u_i, v_i) into groups of sizes $n_1, n_2, \ldots, n_k$. The $u_i(\theta)$ values are defined, as before, as the sequence of x_i values 'corrected' for their group location parameters $\theta_{x_2}, \theta_{x_3}, \ldots, \theta_{x_k}$, and the $y_i(\theta)$ values are obtained similarly.

The discussion of the three-sample case shows how to obtain a test statistic for the more general k-sample case. We calculate Q^*, equation (7.8), by first transforming (u_i, v_i) values to (u_i^*, v_i^*) values, then computing the one-way analysis of variance statistic B^* for the u_i^* and v_i^* sequences separately and finally adding the two statistics; see equation (7.9) and Section 6.4.1. The resulting Q^* has a conditional null distribution under randomization with expectation exactly $2(k-1)$ and under suitable conditions its distribution can be taken as approximation $\chi^2_{2(k-1)}$.

Application of the basic idea after various transformations follows exactly the pattern exhibited in some detail for the three-sample case in Section 7.5, and further details will not be supplied.

Exercises

7.1 In the following table are the results of 24 determinations of X, percentage dry matter in fresh spinach and Y, percentage preserved ascorbic acid after drying at $90°C$; the data are from Hald (1952) p. 904.

X	Y	X	Y
10.0	70.9	12.5	74.2
8.9	74.0	12.3	83.1
8.9	58.6	10.0	66.7
9.2	80.6	10.2	77.2
7.8	69.4	11.2	83.8
10.1	76.0	11.2	67.9
9.0	66.4	10.0	88.9
8.2	50.9	10.7	69.0
9.5	61.9	10.3	69.8
10.8	65.2	12.9	86.0
11.1	77.2	11.8	79.9
11.2	89.6	14.9	88.2

Perform tests of independence of X and Y using the Spearman and the Kendall rank correlation coefficients.

7.2 Using the sign statistics, Section 7.3.1, examine the hypothesis that the medians of X and Y in Exercise 7.1 are 10.5 and 74.0 respectively using the data as given.

7.3 Sketch an outline of a joint 90% confidence region for the population median (θ_1, θ_2) of the joint distribution of X_1 and X_2 in Exercise 7.1, using the given data and the sign statistics.

7.4 The joint distribution of X and Y is uniform on a rectangle which in the (x, y) plane is bounded by the four lines with (slope, intercept) values $(1, 1)$, $(1, -1)$, $(-1, 4)$, $(-1, -4)$.

Calculate the efficacy of the bivariate sign test of location and compare it with the efficacy of a test based on sample means.

7.5 The data given below are extracted from a larger collection reported in Morrison (1967). They are scores on a psychomotor testing device of patients suffering from cancerous lesions. The scores on two days, 1, 2, are shown for each patient. The two groups of results shown are for a control group of size $m = 6$ and a group of size $n = 14$ treated with 25–50 r radiation dosages.

Controls ($m = 6$)		25–50 r		($n = 14$)	
1	2	1	2	1	2
223	242	53	102	32	97
72	81	45	50	38	37
172	214	47	45	66	131
171	191	167	188	210	221
138	204	193	206	167	172
22	22	91	154	23	18
		115	133	234	260

Test the joint equality of Day 1 and Day 2 medians of the two populations represented by these data.

7.6 Assuming symmetry as in Section 7.4 perform a test of equality of location of two bivariate distributions from which the observations in Exercise 7.5 were sampled, using rank statistics.

Appendix A

We saw in Chapter 1 how a 'linearization' technique explained heuristically the asymptotic behaviour of estimators $\hat{\theta}$ defined implicitly as solutions for t of the estimating equation

$$S_n(t) = S_n(\mathbf{X}, t) = 0$$

where $\mathbf{X} = (X_1, \ldots, X_n)$ is a vector of random observations, and where in some contexts '$= 0$' may mean 'change of sign'. To recapitulate, if θ_0 denotes the true value of the unknown parameter θ, then by writing $t = \theta_0 + n^{-1/2}\alpha$ for values in a small neighbourhood of θ_0, it may be possible to expand S_n as

$$S_n(t) = S_n(\theta_0) + \alpha c_n + o(1) \tag{A.1}$$

where $o(1)$ denotes a term which is small in some sense, where S_n has been normalized so that

$$S_n(\theta_0) \xrightarrow{D} N(0, \sigma_0^2) \quad \text{as} \quad n \to \infty \tag{A.2}$$

and where

$$\lim_{n \to \infty} c_n = c_0. \tag{A.3}$$

Then, letting $\hat{\alpha} = n^{1/2}(\hat{\theta} - \theta_0)$, noting that $S_n(\hat{\theta}) = 0$, and ignoring the $o(1)$ term in (A.1) yields

$$\hat{\alpha} = -S_n(\theta_0)/c_n$$

and (A.2) and (A.3) then indicate that

$$\hat{\alpha} = n^{1/2}(\theta - \theta_0) \xrightarrow{D} N(0, \sigma_0^2/c_0^2) \quad \text{as} \quad n \to \infty. \tag{A.4}$$

The equation (A.1) is termed a *linearization representation*. This appendix is concerned with making rigorous the informal reasoning leading to (A.4). Some of the detail which follows is necessarily technical in nature, and assumes a familiarity with probability theory, but some intuitive remarks can be made about the ideas now to be developed.

The argument leading to (A.4) relied upon replacing an arbitrary constant α in (A.1) by the random variable $\hat{\alpha}$. This step can be made justifiable if the precise nature of the smallness of the $o(1)$ term in (A.1) is strong enough. Condition (A.6) below validates the replacement of α by $\hat{\alpha}$, as long as the collection of random variables $\{\hat{\alpha}, n \geq 1\}$ is well-behaved.

The use of linearization techniques has become widespread in asymptotic theory of theoretical statistics (see Brown and Kildea (1979) Jureckova (1969)).

Firstly, rewrite (A.1) as

$$S_n(\theta_0 + n^{-1/2}\alpha) = S_n(\theta_0) + \alpha c_n + U_n(\alpha) \qquad (A.5)$$

and consider the condition:

$$\left.\begin{array}{c} \text{for all finite } M < \infty, \\ \sup_{|\alpha| \leqslant M} |U_n(\alpha)| \xrightarrow{P} 0 \quad \text{as} \quad n \to \infty \end{array}\right\} \qquad (A.6)$$

Proposition A.1. Let (A.2), (A.3) and (A.6) hold, and suppose that $[\hat{\alpha} = n^{1/2}(\hat{\theta} - \theta_0), n \geqslant 1]$ is *tight*, i.e. given $\varepsilon = 0$, there exists K so large that

$$P(|\hat{\alpha}| > K) \leqslant \varepsilon \quad \text{for all} \quad n \geqslant 1. \qquad (A.7)$$

Then (A.4) holds.

Proof. Identifying the M in (A.6) with the K in (A.7) indicates that the remainder term $U_n(\alpha)$ in (A.5) is small in probability when $\hat{\alpha}$ replaces α. Thus U_n may be ignored in the reasoning, outlined previously, which leads to (A.4).

In their turn, the conditions (A.6) and (A.7) are implied by other, simpler conditions. A choice suitable for many applications is given by the following.

Proposition 2. Let (A.2) and (A.3) hold, and also

$$S_n(t) \text{ is monotone increasing in } t \qquad (A.8)$$

and for all fixed α

$$U_n(\alpha) \xrightarrow{P} 0 \quad \text{as} \quad n \to \infty. \qquad (A.9)$$

Then (A.4) holds.

Proof. From Proposition A.1, it is enough to show that (A.8) and (A.9) imply (A.6) and (A.7).

For fixed M and arbitrary ε, choose K so large that $M/K \leqslant \varepsilon$. Let

$$V_{nj} = \sup_{j-1 \leqslant K\alpha \leqslant j} |U_n(\alpha) - U_n\{(j-1)/K\}|$$

where, as in (A.5), $U_n(\alpha) = S_n(\theta_0 + n^{-1/2}\alpha) - S_n(\theta_0) - \alpha c_n$.

Now

$$\sup_{|\alpha| \leqslant M} U_n(\alpha) \leqslant \max_{|j| \leqslant MK} \left\{ \left| U_n\left(\frac{j-1}{K}\right) \right| + V_{nj} \right\}$$

$$\leqslant \max_{|j| \leqslant MK} \left\{ \left| U_n\left(\frac{j-1}{K}\right) \right| \right\} + \max_{|j| \leqslant MK} (V_{nj}) \qquad (A.10)$$

But K is fixed, and $U_n[(j-1)/K] \xrightarrow{P} 0$ for all j, from (A.9). Thus

$$\max_{|j| \leqslant MK} \left\{ \left| U_n\left(\frac{j-1}{K}\right) \right| \right\} \xrightarrow{P} 0 \text{ as } n \to \infty \qquad (A.11)$$

Also, each

$$|V_{nj}| \leqslant S_n\left(\frac{j}{K}\right) - S_n\left(\frac{j-1}{K}\right) + |c_n\varepsilon|$$

from (A.8)

$$\leqslant \left| U_n\left(\frac{j}{K}\right) - U_n\left(\frac{j-1}{K}\right) \right| + 2|c_0|\left(\varepsilon + \frac{1}{K}\right)$$

for n large enough. Thus the limit in probability of each $|V_{nj}|$ is $\leqslant 2|c_0|$ $(\varepsilon + K^{-1})$. But ε and K are arbitrary, so in fact the probability limit is zero. Thus with (A.10) and (A.11), (A.6) is seen to hold.

To establish (A.7), note that the monotonicity of S_n (i.e. (A.8)) means that the event $\{\hat{\alpha} > K\}$ coincides with $\{S_n(\theta_0 + n^{-1/2}K) < 0\}$. But

$$S_n(\theta_0 + n^{-1/2}K) = U_n(K) + Kc_n + S_n(\theta_0).$$

The behaviour of the three terms on the right-hand side is indicated by (A.9), (A.3) and (A.2). Clearly, choice of K large enough ensures that $P(\hat{\alpha} > K) \leqslant \varepsilon$ for all $n \geqslant 1$. $P(\hat{\alpha} < -K)$ is handled in the same way, and (A.7) is established.

Finally, it remains to point out that in most cases, a suitable choice of $\{c_n\}$ enabling (A.3) to hold is just

$$c_n = n^{-1/2}m_n'(\theta_0) \tag{A.12}$$

where $m_n(t) = E\{S_n(t)\}$ is assumed continuously differentiable at $t = \theta_0$. Then, the condition (A.9) is implied by

$$\begin{cases} \text{for all fixed } \alpha, \\ \text{var}\{S_n(\theta_0 + n^{-1/2}\alpha) - S_n(\theta_0)\} \to 0 \text{ as } n \to \infty \end{cases} \tag{A.13}$$

To see this, note that

$$U_n(\alpha) = S_n(\theta_0 + n^{-1/2}\alpha) - S_n(\theta_0) - \alpha n^{-1/2}m_n'(\theta_0).$$

Thus $U_n(\alpha)$ has mean $m_n(\theta_0 + n^{-1/2}\alpha) - m_n(\theta_0) - \alpha n^{-1/2}m_n'(\theta_0)$ which is $o(n^{-1/2})$, and variance which $\to 0$, from (A.13). Thus (A.9) holds.

To summarize, to justify the informal reasoning leading to (A.4), it will be enough in most cases to:

(i) choose c_n according to (A.12) and verify that (A.3) holds;

(ii) check that S_n is correctly normalized so that (A.2) holds;

(iii) check that (A.8) holds; and

(iv) check that (A.13) holds.

Multiparameter linearization theorems can be constructed in exactly the same fashion as the single parameter case considered here. The details can become very involved, of course, but the basic ideas remain simple.

Example A.1 M-estimation of location. Let $X_1, \ldots, X_n$ be i.i.d. random variables, each distributed symmetrically about an unknown parameter θ_0. Let ψ be monotone increasing, differentiable with bounded derivative, and anti-symmetric about zero. The M-estimate of θ_0 based on ψ is $\hat{\theta}$, where $\hat{\theta} = t$ solves $S_n(t) = 0$, with

$$S_n(t) = -n^{-1/2} \sum_{i=1}^{n} \psi(X_i - t).$$

It is easy to check that:

(i) with c_n chosen by (A.12), (A.3) holds and

$$c_0 = E\{\psi'(X_1 - \theta_0)\};$$

(ii) (A.2) holds and $\sigma_0^2 = E\{\psi^2(X_1 - \theta_0)\}$;

(iii) S_n is monotone increasing in t; and

(iv) $\mathrm{var}\{S_n(\theta_0 + n^{-1/2}\alpha) - S_n(\theta_0)\}$

$= \mathrm{var}\{\psi(X_1 - \theta_0) - \psi(X_1 - \theta_0 - n^{-1/2}\alpha)\}$

$\leqslant (n^{-1/2}\alpha C)^2$ where C is an upper bound on ψ',

$\to 0$ as $n \to \infty$; so that (A.13) holds.

Therefore the asymptotic behaviour of M-estimators is given by (A.4); i.e.

$$n^{1/2}(\hat{\theta} - \theta_0) \xrightarrow{D} N\left(0, \frac{E(\psi^2)}{\{E(\psi')\}^2}\right).$$

Example A.2 The sample median. Letting F_n denote the sample distribution function based on random observations $X_1, \ldots, X_n$, whose true distribution function is F with continuous density f, and true median θ_0. The sample median $\hat{\theta}$ solves $S_n(t) = 0$, where

$$S_n(t) = n^{1/2}\{F_n(t) - \tfrac{1}{2}\}.$$

(i) $m_n(t) = n^{1/2}\{F(t) - \tfrac{1}{2}\}$, and c_n given by (A.12) makes (A.3) hold with $c_0 = f(\theta_0)$.

(ii) (A.2) holds and $\sigma_0^2 = 1/4$.

(iii) S_n is monotone increasing in t.

(iv) $\mathrm{var}\{S_n(\theta_0 + n^{-1/2}\alpha) - S_n(\theta_0)\}$ is

$n^{-1} \times$ {the variance of a $\mathrm{Bi}[n, F(\theta_0 + n^{-1/2}\alpha) - F(\theta_0)]$ r.v.},

and $\to 0$ as $n \to \infty$.

Therefore

$$n^{1/2}(\hat{\theta} - \theta_0) \xrightarrow{D} N\left(0, \frac{1}{4f^2(\theta_0)}\right)$$

a well-known result which can be proved more easily by direct means.

In other examples encountered in the main text, the details involved in checking (i)–(iv) are more substantial, but the method works in most cases.

Appendix B

The following notes on asymptotic efficiency are extracted from Cox and Hinkley (1974) pp. 337–338.

Consider any statistic T which has a limiting $N\{\theta, n^{-1}\sigma_T^2(\theta)\}$ distribution, θ being one-dimensional. Then the ML ratio test of $H_0 : \theta = \theta_0$ based on T has a critical region asymptotically equivalent to

$$\{y; |t - \theta_0|\sqrt{n} \geq k^*_{(1/2)\alpha}\sigma_T(\theta_0)\}$$

with significance level approximately α. The large-sample power function, to first order, for $\theta - \theta_0 = \delta n^{-1/2}$ is

$$\Phi\left\{ -k^*_{(1/2)\alpha} - \frac{\delta}{\sigma_T(\theta_0)} \right\} + 1 - \Phi\left\{ k^*_{(1/2)\alpha} - \frac{\delta}{\sigma_T(\theta_0)} \right\} \tag{B.1}$$

which is symmetric about $\delta = 0$.

If we compare two consistent asymptotic normal statistics T_1 and T_2 with limiting variances $n^{-1}\sigma_{T_1}^2(\theta)$ and $n^{-1}\sigma_{T_2}^2(\theta)$, (B.1) shows that they have asymptotically the same power for the same local values of θ if the sample sizes n_1 and n_2 are related by

$$n_1^{-1}\sigma_{T_1}^2(\theta_0) = n_2^{-1}\sigma_{T_2}^2(\theta_0).$$

Thus, if efficiency is measured by relative sample sizes, then the *asymptotic relative efficiency* of T_1 to T_2 is

$$e(T_1 : T_2) = \frac{\sigma_{T_2}^2(\theta_0)}{\sigma_{T_1}^2(\theta_0)}$$

or, equivalently, the ratio of power function curvatures at θ_0. Notice that the measure $e(T_1 : T_2)$ is independent of $k^*_{(1/2)\alpha}$, and hence of the size of the critical regions. The same measure is appropriate for local alternatives when derived from the significance probability point of view.

It is unnecessary to restrict test statistics to be consistent estimates of θ. More generally, we might suppose that T is consistent for some function $\mu(\theta)$ which is a monotone function of θ, at least near to $\theta = \theta_0$. A corresponding general definition of asymptotic relative efficiency, often called Pitman efficiency, relates to statistics T that are asymptotically $N\{\mu_T(\theta), \sigma_T^2(\theta)\}$. The simple calculation leading to (B.1) generalizes immediately to give the asymptotic relative efficiency of two such statistics T_1 and T_2 as

$$e(T_1 : T_2) = \left\{ \frac{\mu'_{T_1}(\theta_0)}{\mu'_{T_2}(\theta_0)} \right\}^2 \left\{ \frac{\sigma^2_{T_2}(\theta_0)}{\sigma^2_{T_1}(\theta_0)} \right\}.$$

Note that the sensitivity measure $\mu'_T(\theta_0)/\sigma_T(\theta_0)$ is the Pitman efficacy introduced in Section 1.6.2. It is invariant under monotonic transformations of T, as is to be expected.

Bibliography

General references

Bradley, J. V. (1968) *Distribution-free Statistical Tests*. Prentice-Hall, Englewood Cliffs, USA.

Cox, D. R. and Hinkley, D. V. (1974) *Theoretical Statistics*. Chapman and Hall, London.

Fisher, R. A. (1966) *The Design of Experiments* 8th Ed. Oliver and Boyd, Edinburgh.

Hollander, M. and Wolfe, D. A. (1973) *Nonparametric Statistical Methods*. Wiley, New York.

Gibbons, J. D. (1971) *Nonparametric Statistical Inference*. McGraw-Hill, New York.

Hajek, J. and Sidak, Z. (1967) *Theory of Rank Tests*. Academic Press, New York.

Kendall, M. G. and Stuart, A. (1961) *The Advanced Theory of Statistics*. Griffin, London.

Lehmann, E. L. (1975) *Nonparametrics: Statistical methods based on Ranks*. Holden-Day, San Francisco.

Puri, M. L. and Sen, P. K. (1971) *Nonparametric Methods in Multivariate Analysis*. Wiley, New York.

Wilks, S. S. (1962) *Mathematical Statistics*. Wiley, New York.

Text references

Adichie, J. N. (1967) Asymptotic efficiency of a class of nonparametric tests for regression parameters. *Ann. Math. Statist*, **38**, 884–893.

Armitage, P. (1971) *Statistical Methods in Medical Research*. Blackwell, London.

Bartels, R. (1980) A nonparametric test for randomness based on ranks. To be published.

Bradley, J. V. (1968) *Distribution-free Statistical Tests*. Prentice Hall, Englewood Cliffs.

Brown, B. M. and Kildea, D. G. (1979) Outlier-detection tests and robust estimators based on signs of residuals. *Comm. in Statist*, **A8**, 257–270.

Brown, G. W. and Mood, A. M. (1951) On median tests for linear hypotheses. *Proceedings of 2nd Berkeley Symposium,* 159–166.

Burr, Irving W. (1974) *Applied Statistical Methods.* McGraw-Hill, New York, p. 405.

Chernoff, H. and Savage, I. R. (1958) Asymptotic normality and efficiency of certain non-parametric tests statistics. *Ann. Math. Statist.,* **29**, 972–994.

Cox, D. R. (1972) Regression models and life-tables. *J. R. Statist. Soc. B.,* **34**, 187–220.

Cox, D. R. and Hinkley, D. V. (1974) *Theoretical Statistics.* Chapman and Hall, London.

David, H. A. (1970) *Order Statistics.* Wiley, New York.

Fisher, R. A. (1966) *The Design of Experiments* 8th Ed. Oliver and Boyd, Edinburgh.

Freireich, E. J. *et al.* (1963) The effect of 6-mercaptopurina on the duration of steroid induced remissions in acute leukemia. *Blood,* **21**, 699–716.

Gehan, E. A. (1965) A generalized Wilcoxon test for arbitrarily single-censored samples. *Biometrika,* **52**, 203–224.

Gibbons, J. D. (1971) *Nonparametric Statistical Inference.* McGraw-Hill, New York.

Greenwood, M. (1926) *The 'errors of sampling' of survivorship tables.* Reports on Public Health and Medical Subjects, No. 33, Appendix 1, H. M. Stationery Office, London.

Hájek, J. and Sidák, Z. (1967) *Theory of Rank Tests.* Academic Press, New York.

Huber, P. J. (1972) Robust statistics: a review. *Ann. Math. Statist.,* **43**, 1041–1067.

Hodges, J. L. Jnr. and Lehmann, E. L. (1963) Estimates of location based on rank tests. *Ann. Math. Statist.* **34**, 598–611.

Johnson, N. L. and Leone, F. C. (1964) *Statistics and Experimental Design in Engineering and the Physical Sciences* Vol 1. Wiley, New York.

Jureckova, J. (1969) Asymptotic linearity of a rank statistic in regression parameter. *Ann. Math. Statist.,* **40**, 1889–1900.

Kaplan, E. L. and Meier, P. (1958) Nonparametric estimation from incomplete observations. *J. Am. Statist. Assoc.,* **53**, 457–481.

Kendall, M. G. (1955) *Rank Correlation Methods* 2nd Ed. Hafner, New York.

Kendall, M. G. and Stuart, A. (1961) *The Advanced Theory of Statistics* Vol. II. Griffin, London.

Kerrich, J. E. (1955) Confidence intervals associated with a line fitted by least squares. *Statistica Neerlandia,* 125–129.

Lehmann, E. L. (1975) *Nonparametrics: Statistical Methods based on Ranks.* Holden-Day, San Francisco.

Maritz, J. S. and Jarrett, R. G. (1978) A note on estimating the variance of the sample median. *J. Am. Statist. Assoc.,* **73**, 194–196.

Mood, A. M. (1954) On the asymptotic efficiency of certain nonparametric two-sample tests. *Ann. Math. Statist.*, **25**, 514–522.

Morrison, D. F. (1967) *Multivariate Statistical Methods.* McGraw-Hill, New York.

Owen, D. B. (1962) *Handbook of Statistical Tables.* Addison-Wesley, Reading, Mass.

Parzen, E. (1962) On estimation of a probability density function and mode. *Ann. Math. Statist.*, **33**, 1065–1076.

Pike, M. C. and Roe, F. J. C. (1963) An acturial method of analysis of an experiment in two-stage carcinogenesis. *Brit. J. Cancer*, **17**, 605–10.

Pitman, E. J. G. (1937) Significance tests which may be applied to samples from any population. *I. Suppl. J. R. Statist. Soc.*, **4**, 119–130.

Puri, M. L. and Sen, P. K. (1971) *Nonparametric Methods in Multivariate Analysis.* Wiley, New York.

Rosenblatt, M. (1971) Curve estimates. *Ann. Math. Statist.*, **42**, 1815–1842.

Terry, M. E. (1952) Some rank order tests which are most powerful against specific alternatives. *Ann. Math. Statist.*, **23**, 346–366.

Theil, H. (1950) A rank invariant method of linear and polynomial regression analysis. *Proc. Kon. Ned. Akad. v Wetensch. A*, **53**, 386–392.

Wald, A. and Wolfowitz, J. (1944) Statistical tests based on permutations of the observations. *Ann. Math. Statist.*, **15**, 359–372.

Westenberg, J. (1948) Significance test for median and interquartile range in samples from continuous population of any form. *Proc. Kon. Ned. Akad. v. Wetensch.*, **51**, 252–261.

Wilcoxon, F. (1945) Individual comparisons by ranking methods. *Biometrics*, **1**, 80–83.

Wilks, S. S (1962) *Mathematical Statistics.* Wiley, New York.

Subject index